AF179694

Hazardous Waste Incineration and Human Health

Authors

Curtis C. Travis, Ph.D.
Director
Office of Risk Analysis
Health and Safety Research Division
Oak Ridge National Laboratory
Oak Ridge, Tennessee

S. Chrystal Cook
Project Leader, ULTRA Publications
Technical Publications Specialist
SE-Retail Division
NCR Corporation
Atlanta, Georgia

CRC Press
Boca Raton Ann Arbor Boston

Library of Congress Cataloging-in-Publication Data

Travis, C. C.
Hazardous waste incineration and human health/authors Curtis C.
Travis, S. Chrystal Cook. p. cm.
Bibliography: p.
Includes index.
ISBN 0-8493-6754-9
1. Hazardous wastes — Incineration — Health aspects — United States.
2. Health risk assessment — United States. I. Cook, S. Chrystal.
II. Title.
RA578.H38T7 1989 88-14585
363.7'28—dc19

This book represents information obtained from authentic and highly regarded sources. Reprinted material is quoted with permission, and sources are indicated. A wide variety of references are listed. Every reasonable effort has been made to give reliable data and information, but the author and the publisher cannot assume responsibility for the validity of all materials or for the consequences of their use.

All rights reserved. This book, or any parts thereof, may not be reproduced in any form without written consent from the publisher.

Direct all inquiries to CRC Press, Inc., 2000 Corporate Blvd., N.W., Boca Raton, Florida, 33431.

© 1989 by CRC Press, Inc.
Second Printing, 1990

International Standard Book Number 0-8493-6754-9

Library of Congress Card Number 88-14585
Printed in the United States

PREFACE

A serious environmental problem confronting industrialized nations today is how to dispose of the large quantities of hazardous waste chemicals generated by product manufacturers and users. It is estimated that there are about 15,000 industrial generators of hazardous waste in the U.S. and that collectively they produce more than 264 million metric tons annually.[1] Only an estimated 10% of this hazardous material is being disposed of in an environmentally safe manner.[2]

Currently hazardous waste is handled in one of three ways: treatment, storage, or disposal. Treatment strategies include biological decomposition and high-temperature incineration. Storage consists of methods such as holding tanks and other above-ground containers. Disposal methods include landfilling and surface impoundment. Some potential problems incurred through these hazardous waste management methods are described below.

Many U.S. disposal sites (landfill and impoundment) are located in sandy or gravelly terrain only a few feet above groundwater, and frequently in close proximity to shallow drinking-water wells. Fred C. Hart Associates, Inc.[3] studied 929 waste disposal sites throughout the U.S. Typical sites were not constructed with plastic or clay liners to retard leakage, and inactive sites had invariably received inadequate closure such as capping with clay or dirt. In most cases, facilities were located in permeable soils within 10 feet of groundwater and within 100 feet of a surface water body.

Once organic chemicals enter the environment, they frequently find their way into drinking water supplies via seepage into aquifers.[4] Data from several sources indicate contamination of drinking-water wells by toxic organic chemicals in over 40 states. For example, in a survey of over 1,000 New Jersey wells, Page[5] found the probability of well contamination by the more prevalent chemical species to be: 1,1,1-trichloroethane, 78% chloroform, 64%; carbon tetrachloride, 64%; 1,1,2-trichloroethylene, 58%; trans-dichloroethylene, 51%; BHC-beta, 50%; and 100% for the nine heavy metals — arsenic, beryllium, cadmium, copper, chromium, nickel, lead, selenium, and zinc.

The public is becoming increasingly aware of the pollution of drinking water by synthetic organic chemicals in addition to other forms of environmental pollution. There is a growing movement to embrace the concept of prevention of contamination by permanent destruction of hazardous waste through incineration. Numerous incidents have proven that long-term containment of hazardous wastes is difficult at best, and that restoration of a contaminated aquifer is impossible for both technical and economic reasons. Permanent destruction, while not inexpensive, may prove to be the most sound long-term economic investment.

However, no hazardous waste management option is risk free; incineration is no exception. As hazardous waste streams are incinerated, most of the toxic components are thermally destroyed or removed from the waste stream in bottom ash. However, a small fraction of the organic constituents, in the average range of 0.01%, are released up the incinerator stack to the atmosphere. Moreover, up to 35% of metals in a hazardous waste stream can be emitted to the atmosphere during incineration. Many of these organics and metals are carcinogenic and may have adverse health effects on populations residing within the vicinity of the incinerator.

Risk analysis can be defined as "the collection and examination of technical and scientific data to measure the probability and severity of adverse effects".[6] Using state-of-the-art risk assessment procedures, we hope to clarify the nature and magnitude of the health risk resulting from hazardous waste incineration. Our intent is that the information and interpretation presented herein be useful to both layman and scientist, to those whose activities are regulated and those who make regulatory decisions.

Chapter 1 introduces some of the complexities involved in defining, measuring, and regulating the nation's hazardous waste. Questions addressed include: What is hazardous waste? Who generates hazardous waste? Where is most of the nation's hazardous waste produced? What type of hazardous waste, and how much, is incinerated annually? And what chemicals are most often and abundantly detected in incinerated waste?

Chapter 2 provides brief descriptions of the hazardous waste incineration process and the presently available incinerator types. Also noted are thermal destruction technologies that may be available to the industrial community during the next five years. We proceed with an outline of the EPA system for testing incinerator efficiency and issuing operating permits, and conclude the chapter with an overview of feasibility for some nonthermal hazardous waste management options.

Chapter 3 summarizes the data base from incinerator test burns conducted by or for the EPA. The importance of the various measurements as well as the numbers are presented. These measurements, obtained during incinerator operation, are the basis for implementation of the risk assessment process.

Chapter 4 contains a description of the four components of risk analysis, provides sample calculations of both carcinogenic and noncarcinogenic health risk estimates, and discusses the predictive methodology employed in quantitative risk assessment for hazardous waste incinerators. The foci of the procedures presented are inhalation and ingestion exposures. (Surface water, groundwater, and aquatic food chain exposures are not considered in our risk estimates.)

Chapter 5 begins the series of sections which present risk estimates for exposures to hazardous waste incinerator emissions. Inhalation exposure to organic stack emissions from the firing of two generic waste streams is the topic in this chapter; fugitive emissions are not included in the exposure assessment. Population and individual exposures are analyzed for both carcinogenic and noncarcinogenic health risks.

Chapter 6 is a discussion of inhalation exposure to incinerator stack releases of heavy metals. (Again, the impact of fugitive emissions is not considered in the estimations.) The first part of this chapter deals with health effects estimated from measurements of several metals taken during incinerator studies; the second half estimates health effects from exposure to lead, cadmium, and chromium.

Chapter 7 introduces the potential of incinerator design features and/or chemical reactions to produce and release hazardous compounds not present in the original waste burned. Health risk estimates are calculated for these little-understood compounds that are referred to as products of incomplete (or poor) combustion. The second part of the chapter introduces dioxins and furans, two related families of the most toxic compounds known to man, which are considered to be products of incomplete combustion.

Chapter 8 introduces the topic of polychlorinated biphenyl compounds. Although production of these chemicals was banned by Congress in 1979, the existing contaminated material must be disposed of safely. This chapter presents health risk estimates for inhalation exposure to incinerator releases from combustion of polychlorinated biphenyls in both hypothetical and trial burn situations.

Chapter 9 addresses ingestion exposure to incinerator releases through the terrestrial food chain pathway. Three hypothetical incinerator sites were chosen for this assessment. Foodstuffs used in the analysis are vegetables, grains, milk, and beef. Health risk estimates for the ingestion pathway are then compared to inhalation risk in the vicinity of the same facilities.

REFERENCES

1. **Office of Technology Assessment (OTA),** *Technologies and Management Strategies for Hazardous Waste Control,* U.S. Government Printing Office, Washington, D.C., 1983.
2. **Fawcett, H. H.,** *Hazardous and Toxic Materials: Safe Handling and Disposal,* Wiley-Interscience, New York, 1984.
3. **Hart, F. C., (Fred C. Hart Associates, Inc.),** *Impact of Burning of Hazardous Waste in Boilers,* SCA Chemical Services, Inc., Boston, MA, August 1982.
4. **Travis, C. C. and Etnier, E. L.,** *Groundwater Pollution: Environmental and Legal Problems,* AAAS Selected Symposium No. 9, Westview Press, Boulder, CO, 1984.
5. **Page, G. W.,** Comparison of groundwater and surface water for patterns and levels of contamination by toxic substances, *Environ. Sci. Technol.,* 15(12), 1475, 1981.
6. **Conway, R. A., Ed.,** *Environmental Risk Analysis for Chemicals,* Nostrand Reinhold, New York, 1982.

THE AUTHORS

Curtis C. Travis, Ph.D., is Director of the Office of Risk Analysis at Oak Ridge National Laboratory (ORNL), Oak Ridge, Tennessee. ORNL is operated by Martin Marietta Energy Systems, Inc.

Dr. Travis obtained his training at the University of California, Davis, receiving his Ph.D. degree in Applied Mathematics in 1971. He received his M.A. degree in Biomathematics in 1967 and his B.A. degree in Mathematics in 1966 from the California State University, Fresno. He served as Research Engineer at the Jet Propulsion Laboratory of the California Institute of Technology, as Assistant Professor of Mathematics at Vanderbilt University and the University of Tennessee, and as a member of the research staff of ORNL since 1976. It was in 1984 that he assumed his present position.

Dr. Travis is a member of the Society for Risk Analysis and is Editor-in-Chief of *Risk Analysis: An International Journal.* He is a member of the Science Advisory Board of the National Center for Toxicological Research, Chairman of the Office of Science and Technology Policy Task Force on Risk Analysis, and has served on numerous national committees. He is Course Director for the Society for Risk Analysis Course on Carcinogenic Risk Analysis.

Dr. Travis has authored more than 150 papers and has been the author or co-author of 6 books. He has given more than 50 invited presentations on the subject of risk assessment in the last 3 years. His current research interests relate to: (1) the development and application of physiologically-based pharmacokinetic and pharmacodynamic models for use in risk analysis, (2) evaluation of the scientific bases for assumptions used by Federal Agencies in assessing risk, and (3) development of new methodologies in the areas of health assessments, hazardous waste site characterization, and environmental health research.

S. Chrystal Cook is a Technical Publications Specialist and Project Leader for ULTRA Publications in the Systems Engineering-Retail Division of NCR Corporation in Atlanta, Georgia.

Ms. Cook obtained her undergraduate training as a writer at the University of Tennessee at Knoxville, receiving her B.A. in 1984. She has done graduate work in Technical Communications on academic fellowship at Wright State University in Dayton, and is currently pursuing graduate research in Atlanta.

During her work at Oak Ridge National Laboratory, Oak Ridge, Tennessee, as a technical writing consultant to the Office of Risk Analysis in the Health and Safety Research Division, she published papers and book reviews in the risk analysis field, specifically concerning the relationship of toxic chemical incineration to human health.

Ms. Cook is a member of the International Association of Business Communicators, the Society for Technical Communication, and has been named to the following honor societies: Phi Beta Kappa, Phi Kappa Phi, Golden Key, Gamma Beta Phi, and Phi Eta Sigma.

Since 1986, her work has involved information product development and publication of reference and operational documentation for computer technology at NCR Corporation.

TABLE OF CONTENTS

Chapter 1

COMPOSITION OF HAZARDOUS WASTE

I. INTRODUCTION

In this chapter we address the question "What is a hazardous waste?" Under current Federal regulations, there are several ways to define or identify hazardous waste streams and/or the individual constituents of a given stream. These regulations are often difficult to understand or implement by the regulated community. Beginning with the Congressional Act dealing specifically with hazardous waste, we offer a brief answer.

II. RESOURCE CONSERVATION AND RECOVERY ACT

In 1976, the U.S. Congress passed the Resource Conservation and Recovery Act (RCRA). Under Subtitle C of RCRA, a basic regulatory program for national hazardous waste management was established.[1] This act was designed to regulate hazardous waste from the point of generation to final disposal and includes the following components: (1) identification and listing of hazardous waste; (2) a national system to track waste; (3) standards for waste management facility operations; and (4) a permit issue system for hazardous material treatment, storage, and disposal sites. Although all hazardous waste activities in the U.S. are subject to Federal regulation under RCRA, individual states providing regulatory programs as stringent, comprehensive, and effective as the RCRA framework have the option to waive the Federal statutes and exercise primary administration and enforcement authority over hazardous waste.

In passing RCRA, Congress recognized the need for special expertise and accordingly designated the Environmental Protection Agency (EPA) as the authority "to define and identify hazardous waste and its characteristics and to develop a comprehensive system for the control of hazardous waste management and disposal".[1] The EPA received this license from Congress with the broad directive to fulfill one primary hazardous waste regulatory goal: the protection of human health and the environment. By January 26, 1983, the basic Federal regulatory program for hazardous waste was set, although EPA states that the program is far from complete. In 1983, EPA estimated that full implementation of RCRA, through issuance of detailed technical standards and permits for existing hazardous waste facilities and the approval of state programs, could take 5 to 7 more years.[1]

Table 1 is an overview of the Federal statutes, under enforcement jurisdiction of the EPA and other government agencies, which pertain to hazardous and toxic substances.

III. DEFINITION OF HAZARDOUS WASTE

The issue of defining hazardous waste, for the purpose of Federal and state regulation in the interest of health and environmental safety, is both complex and controversial. There is disagreement concerning the accuracy and adequacy of the Federal hazardous waste definition, which leads to the varying estimates of the quantities and types generated throughout the U.S. In classifying waste, "it is not sufficient to merely use information on the most hazardous constituent, or the one present in the largest amount. . . . There is currently no standard procedure to describe the hazard level for a physically and chemically complex waste".[1]

The hazardous waste universe defined by Federal law begins with the RCRA Subtitle C identification of a solid waste. RCRA defines solid waste as "any garbage, refuse, sludge

Table 1
MAJOR LAWS CONTROLLING TOXIC SUBSTANCES[2]

Environmental Protection Agency

Toxic Substances Control Act of 1976 — Places heavy reporting burden on industry; the EPA can demand premarket testing of some chemicals

Resource Conservation and Recovery Act of 1976 — Provides for regulation of disposal of toxic and other hazardous wastes

Safe Drinking Water Act of 1975 — Provides for regulation of toxic substances in public drinking water supplies

Water Pollution Control Act Amendments of 1972 and 1977 — Provides for regulation of discharges of hazardous effluents into the nation's waterways and the ocean within a 3-mi limit; cleanup of hazardous spills on land and in water

Marine Protection, Research, and Sanctuaries Act of 1972 — Provides for regulation of ocean dumping from 3 to 12 mi at sea (enactment of 200-mi limit in spring 1977 may extend this)

Clean Air Act Amendments of 1970 — Allows the EPA to set national emission standards for hazardous air pollutants

Federal Insecticide, Fungicide, and Rodenticide Act of 1972 — Covers the sale and use of economic poisons and foodstuff treated with such poisons; counterpart legislation exists for enforcement by the USDA and the FDA

Food and Drug Administration

Federal Food, Drug, and Cosmetic Act of 1906 (amended 1938 and 1962) — Bars any detectable amounts of carcinogens in foods and cosmetics; specifically exempts hair dyes from enforcement provisions

Occupational Safety and Health Administration

Occupational Safety and Health Act of 1970 — Regulates hazardous materials in the workplace or in products bought for use in the workplace

Consumer Product Safety Commission

Consumer Product Safety Act of 1972 — Specifically excludes tobacco, drugs, foods, and cosmetics from regulations; created commission to enforce, among other things, earlier statutes

Federal Hazardous Substances Act of 1927 (amended 1976) — Provides for regulation of flammable, corrosive, allergenic or toxic materials in consumer products

Flammable Fabrics Act of 1953 (as amended) — Concerns clothing as well as fabrics used for other purposes in the home

Poison Prevention Packaging Act — Produced childproof caps

Department of Transportation

Oil Pollution Act of 1961 — Concerns oil spills and chemical spills from ships

Dangerous Cargo Act, Tank Vessel Act, Ports and Waterways Safety Act 1972, and Pipeline Safety Act — Concern safety for various railroad and truck transportation

From Fawcett, H. H., *Hazardous and Toxic Materials: Safe Handling and Disposal,* John Wiley & Sons/Interscience, New York, 1984. With permission.

. . . and other discarded material, including solid, liquid, semisolid, or contained gaseous material resulting from industrial, commercial, mining, and agricultural operations, and from community activities . . . ."[1]

Any person generating a solid waste is responsible for determining whether that solid waste is a hazardous waste. RCRA-regulated hazardous waste is defined as a solid waste (or mixture of solid wastes) which may "cause or significantly contribute to an increase in mortality or an increase in serious irreversible or incapacitating reversible illness"; or poses "a substantial present or potential hazard to human health or the environment when it is improperly treated, stored, transported, disposed of, or otherwise managed".[3] It is important to emphasize here that wastes (their individual chemical components) are not hazardous in and of themselves; rather, the levels of hazard (or safety) are dependent upon the proper utilization of waste management techniques.[2]

After the RCRA definitions of solid waste and hazardous waste were established, Congress directed the EPA to follow a two-step method in order to identify hazardous wastes.[4] A solid waste is identified as hazardous if it is (1) listed by the EPA on one of four hazardous waste lists or if it (2) exhibits one or more of the four EPA-defined hazardous waste characteristics.

A. Step One: Hazardous Waste Lists

Four separate hazardous waste lists containing more than 400 substances are published in the *Code of Federal Regulations (CFR)*:[5]

1. Hazardous wastes from nonspecific sources listed as "F" wastes[6]
2. Hazardous wastes from specific industrial processes listed as "K" wastes[7]
3. Discarded commercial chemical products that are determined acutely hazardous and listed as "P" wastes[8]
4. Discarded commercial chemical products that are determined toxic and listed as "U" wastes[8]

The EPA Appendix list of hazardous constituents, which is also found in 40 CFR, Part 261, is not a separate hazardous waste list. Rather, the substances included in Appendix are those which, if found in a solid water, "will cause the EPA to consider listing that waste as hazardous".[4] Chemicals are included in the Appendix list only if "they have been shown in scientific studies to have toxic, carcinogenic, mutagenic, or teratogenic effects on humans or other life forms".[4] These effects are defined as follows:

- Toxic — presenting an unreasonable risk to human health or the environment
- Carcinogenic — having cancer-causing potential
- Mutagenic — capable of producing mutation in genetic material
- Teratogenic — affecting the genetic characteristics of an organism so as to cause malformations, monstrosities, or serious deviations from normal growth

B. Step Two: Characteristic Waste

Waste components not previously listed by the EPA can be determined hazardous or nonhazardous when tested according to the following Federally defined criteria:[1]

- Ignitability — the characteristic of posing a fire hazard during routine waste management
- Corrosivity — the ability to corrode waste storage containers or to dissolve toxic components of other wastes
- Reactivity — the tendency to explode, react violently when mixed with water, or generate toxic gasses under normal management conditions
- EP Toxicity — the presence of certain toxic materials at levels higher than those specified in Code of Federal Regulations (1985)

The EPA has limited the criteria used to define hazardous waste to the above four, since it is believed that these characteristics are the only ones for which test protocols are readily available to the regulated community.[9] Wastes with components exhibiting one or more of these characteristics are determined hazardous and assigned an EPA hazardous waste number and code. (EPA has plans to expand the basic set of characteristics to include radioactivity, unnatural genetic activity, bioaccumulation, and toxicity to aquatic organisms, terrestrial plants, and humans [chronic exposure], as regulatory testing protocols for these characteristics become available.)[9] Brief descriptions of the four hazardous waste testing characteristics follow.

1. Ignitability

Ignitability criteria characterize wastes that, under routine handling, could start fires or exacerbate fires once started. Ignitable waste can cause damage directly through heat and smoke or indirectly by providing the means by which other hazardous materials disperse (i.e., convection in flames produced by ignitable wastes). Otherwise harmless substances, such as plastics can become hazardous by emitting toxic fumes during combustion caused by an ignitable substance.[10]

The EPA has determined that a solid waste exhibits the characteristic of ignitability of a representative sample of that waste:[11]

1. Is a liquid containing less than 50% water and less than 24% alcohol by volume and has a flash point (FP) less than 60°C (140°F) (The FP is the "lowest temperature at which vapors above a volatile combustible substance ignite in air when exposed to flame".[12])
2. Is a nonliquid capable of causing fire through friction, absorption of moisture, or spontaneous chemical changes under standard temperature and pressure; when ignited, burns so vigorously and persistently that a hazard is created
3. Is an ignitable compressed gas as defined by the Department of Transportation (DOT)
4. Is an oxidizer as defined by the Department of Transportation (DOT) regulations

Any solid waste that exhibits one or more of these ignitable properties and is not previously identified by EPA listing is determined a hazardous waste and can be classified under the EPA hazardous waste number D001. The EPA hazard code for ignitability is "I".[11]

2. Corrosivity

Corrosive wastes present two primary health concerns. First is the safety of waste handlers whose tissues may be damaged by the corrosive action of waste components. Second, corrosive wastes may corrode their storage containers and leak out to the environment, causing further damage.[10] Proper labeling to insure cautious handling is the key variable in preventing unnecessary corrosion risk.

A solid waste exhibits the characteristic of corrosivity if a representative sample of the waste:[13]

1. Is aqueous (the primary constituent of an aqueous waste is water) and has a pH less than or equal to 2 or greater than or equal to 12.5
2. Is liquid and corrodes steel at a rate greater than 6.35 mm (0.250 in.) per year under a test temperature of 55°C (130°F):

Any solid waste that exhibits either of the corrosive properties and is not previously identified in one of the four EPA hazardous waste lists, is determined a hazardous waste and can be classified under the EPA hazardous waste number D002. The EPA hazard code for corrosives is "C".[13]

3. Reactivity

Reactive wastes are those that are extremely unstable under normal conditions, with a tendency to explode or give off dangerous gases. Reactivity of a waste cannot be predicted from a particular structural pattern or chemical composition. Standard methods to test for reactivity currently do not exist; in order to describe reactive wastes, it is possible only to list their physical peculiarities.[10]

A solid waste exhibits the characteristic of reactivity, according to the EPA definition, if a representative sample of the waste:

1. Is typically unstable and readily undergoes violent change without detonating
2. Reacts violently with water
3. Forms potentially explosive mixtures with water
4. When mixed with water, generates toxic gases, vapors, or fumes in a quantity sufficient to present a danger to human health or the environment
5. Is a cyanide- or sulfide-bearing waste which, when exposed to pH conditions between 2 and 12.5, can generate toxic gases, vapors, or fumes in a quantity large enough to pose a threat to humans or the environment
6. Is capable of detonation or explosive reaction if subjected to a strong initiating source or if heated under confinement
7. Is readily capable of detonation, explosive decomposition, or explosive reaction at standard temperature and pressure
8. Is a forbidden explosive, or a Class A or Class B explosive, as defined in DOT regulations

A solid waste that exhibits one or more of the above properties of reactivity, but is not previously listed as a hazardous waste by the EPA, is classified as a hazardous waste by the EPA number D003.[14]

4. EP Toxicity

The fourth major characteristic by which a hazardous waste can be identified is EP toxicity. This characteristic is a measure of the likelihood that the waste will leach out toxic chemicals. To be classified as a hazardous waste by EP toxicity standards, a specified extraction procedure must yield an extract from the waste with concentrations equal to or above the following designated amounts:

Contaminant	**Concentration (mg/l)**
Arsenic	5.0
Barium	100.0
Cadmium	1.0
Chromium	5.0
Lead	5.0
Mercury	0.2
Selenium	1.0
Silver	5.0
Endrin	0.2
Lindane	0.4
Methoxychlor	10.0
Toxaphene	0.5
2,4-D	10.0
2,4,5-TP	1.0

Any solid waste exhibiting the above contaminant concentration(s) that is not previously listed by the EPA as a hazardous waste is classified a hazardous waste and assigned the

Table 2
HAZARDOUS WASTE IDENTIFICATION[5]

Waste type	EPA hazardous waste number
Listed wastes	
Nonspecific source waste	Fxxx
Specific source waste	Kxxx
Acute hazardous waste	Pxxx
Toxic waste	Uxxx
Characteristic wastes	
Ignitable waste (I)	D001
Corrosive waste (C)	D002
Reactive waste	D003
EP toxic waste (E)	D004—D017

EPA hazardous waste numbers D004 through D017. The hazard code for EP toxic wastes is "E".[15]

Table 2 summarizes the ways in which hazardous wastes are currently identified by the EPA and the regulated community.

Any waste classified as hazardous, whether by the EPA listing or testing by the four characteristics, is under Federal regulation unless it has been delisted.[1] There are two possible approaches to the delisting of a waste. In the first method, a petition seeking delisting of a waste from a specific generator must demonstrate that the waste is not hazardous under the individual circumstances. The second type of petition seeks to remove a waste from hazardous classification on the premise that EPA erroneously included the waste (i.e., the waste does not meet any of the specified criteria). The delisting process takes many factors into consideration regarding unique waste streams and generators and can be an incentive for generators to change their processes or products so they will not be labeled hazardous.

IV. QUANTITIES OF HAZARDOUS WASTE

Determination of the quantity of hazardous waste generated annually is complicated by controversial issues involved with RCRA specifications including: (1) Federal vs. state definitions of the hazardous waste universe (a large number of exemptions are included under the Federal program) and (2) lack of regulation for hazardous wastes that result from products used in the home (i.e., cleaning solutions and pesticides).[2] The result is widely varying estimated quantities which further complicate regulatory tasks.

A. Estimates

According to statistics from the Office of Technology Assessment (OTA), approximately 255 to 275 million t of hazardous waste are generated annually.[1] National survey data collected for the EPA confirmed this estimate, indicating that 264 million t (71 billion gal) of hazardous waste were generated during 1981.[16] Althouth this estimate contained mixtures of hazardous and nonhazardous materials, the combined waste streams were RCRA-defined hazardous substances and were necessarily included in the total. This survey did not, however, estimate the quantity of all hazardous waste produced in 1981; the 264 million t included only the hazardous waste that would be managed in Federally regulated treatment, storage, and disposal processes.

B. Issues

A major inadequacy of the RCRA program is the large number of exemptions under the

Table 3
EXEMPTIONS AND EXCLUSIONS FROM THE FEDERALLLY DEFINED UNIVERSE OF HAZARDOUS WASTE[1]

Exclusions from the statutory definition of solid waste

Solid or dissolved materials in domestic sewage
Solid or dissolved materials in irrigation return flows
Industrial discharges that are point sources subject to National Pollutant Discharge Elimination System (NPDES) permits under Section 402 of the Clean Water Act
Source, special nuclear, or byproduct material defined under the Atomic Energy Act of 1954, as amended

Exclusions by rule from the definition of solid waste

Statutory exclusions above
Waste burned as fuel for purposes of recovering usable energy
In-situ mining wastes not removed from the ground

Temporary statutory exclusions from the definition of hazardous waste

Drilling fluids produced waters, and other wastes associated with the exploration, development, or production of crude oil, natural gas, or geothermal energy
Fly ash waste, bottom ash waste, slag waste, and flue gas emission control waste generated primarily from combustion of coal or other fossil fuels
Solid waste from the extraction, beneficiation, and processing of ores and minerals, including phosphate rock and the overburden from the mining of uranium ore
Cement kiln dust waste

Solid wastes excluded from the definition of hazardous wastes in EPA regulations

Household waste
Agricultural- and livestock-raising wastes used as fertilizers
Mining overburden returned to the minesite
Temporary statutory exclusions above
Certain wastes containing exclusively (or almost exclusively) trivalent chromium from leather-tanning and fishing industries, shoe-manufacturing and other leather product industries, and wastewater treatment sludge from production of TiO_2 pigment from chromium-bearing ores by the chloride process (if not hazardous under any other provision except failure of the EP toxicity test for chromium)
Solid waste from coal mining and processing
Arsenical-treated wood or wood products which: (1) fail the test for EP toxicity and (2) are discarded by persons using the wood or wood products for its intended-end use (unless the waste meets other tests for hazardous waste)
Any waste, sludge, or residue for hazardous waste treatment that is no longer hazardous because it no longer displays a characteristic of hazardous waste
Hazardous waste generated in a produce or raw material storage tank, transport vehicle, or in a closed manufacturing process unit or waste treatment unit before it exits from or is removed from the unit
Samples of solid waste or of water, soil, or air collected solely for testing, subject to special handling requirements to qualify for this exemption)
A delisted solid waste or sludge or residue from treatment of a delisted hazardous waste (provided that it does not exhibit any characteristics of hazardous waste)

Federal definition of hazardous waste (Table 3). These exemptions limit Federal regulations in contrast with the broader controls enforced in some states. In fact, Federal programs control only about one fourth of the annual amount of hazardous waste generated.[1] Although much of the Federally exempted waste has relatively low levels of hazard, some of the excluded substances are substantially threatening to human health and the environment. Wastes produced by small quantity generators, for example, are not under Federal regulation, although the waste produced may be identical in composition to that from larger generators.

The annual total of hazardous waste that is not controlled by Congress and the EPA is

Table 4
EXAMPLES OF EXEMPTIONS FROM FEDERAL REGULATION AS HAZARDOUS WASTE[1]

Waste type	Estimated annual generation (million t)	Possible hazard
Fly and bottom ash from burning fossil fuels	66	Trace toxic metals
Fuels gas emission control waste	Unknown	Toxic organics and inorganics
Mining waste, including radioactive waste	2100	Toxic metals, acidity, radioactivity
Domestic sewage discharged into publicly owned treatment works	5	Uncertain, toxic metals likely
Cement kiln dust	12	Alkalinity, toxic metals
Gas and oil drilling muds and production waste, geothermal energy waste	Unknown	Alkalinity, toxic metals, toxic organics, salinity
NPDES-permitted industrial discharge	Unknown	Toxic organics, heavy metals
Irrigation return flows	Unknown	Pesticides, fertilizers
Wastes burned as fuels	19	Unburned toxic organics
Waste oil	Unknown	Toxic organics, toxic metals
Infectious waste	Unknown	Infectious materials
Small volume generators	2.7—4.0	Possibly any hazardous waste
Agricultural waste	Unknown	Variable
Waste exempted under delisting petitions[a]	Unknown	Presumably insignificant[a]
Deferred regulations	Unknown	Unknown
EPA deregulation	Unknown	Presumably insignificant
Toxicity test exemptions[b]	Unknown	Organics
Recycled waste[c]	Unknown	Improper application of various materials

[a] Wastes may be delisted on the basis of a petition that is concerned with the constituent(s) which have determined the original listing; however, other hazardous constituents may be present which have previously been unrecognized administratively.

[b] Wastes not identified as toxic by the EPA extraction procedure test and not otherwise listed by EPA.

[c] Legitimate recycling is exempt from RCRA regulations except for storage; however, there have been numerous incidents (e.g., the dioxin case in Missouri) involving recycled materials which are still hazardous.

several hundred million t.[1] Table 4 gives examples of the estimated annual generation for some of the Federally exempted wastes. Table 5 compares the Federal and state estimates of annual hazardous waste quantities, specifying those hazardous materials governed by state law that are not included in the Federally defined hazardous waste universe.

A second inadequacy of the RCRA specifications is that hazardous wastes resulting from consumer products, such as cleaning solutions and pesticides, are not regulated in the home.[2] While these waste quantities are small in proportion to industrial output of hazardous waste, their effect becomes significant immediately upon entering a landfill, water stream, or incinerator. Once a hazardous waste is collected, stored, transported, treated, and/or disposed of, it becomes subject to local, state, and Federal law, although no regulatory statutes exist for the previous handling and use of that material in the home.

V. GENERATORS OF HAZARDOUS WASTE

A generator of hazardous waste is ''any person, by site, whose act or process produces hazardous waste or whose act first causes a hazardous waste to be subject to regulation''.[3] In the EPA survey conducted in the fall of 1982 and the spring of 1983, 60,000 companies indicated that they either presently generated hazardous waste or might become generators in the immediate future. Of the reporting firms, 15,000 responded that they treated, stored, and/or disposed of RCRA-regulated hazardous waste.[16]

Table 5
HAZARDOUS WASTE GENERATION ESTIMATES BY EPA AND THE STATES[1]

	Quantity (t)		Universe	
State	EPA estimate	State estimate	Same as EPA	State additions to EPA hazardous waste definition
Alabama	730,000	265,680		PCBs
Alaska	130,000	360		PCBs
Arizona	160,000	4,280,000		PCBs, waste oil
Arkansas	370,000	No data		PCBs, waste oil
California	2,630,000	15,000,000		Approximately 4 million t is oilfield waste; also includes mining waste, small generators, PCBs
Colorado	180,000	775,490		PCBs
Connecticut	610,000	102,000	x	Extrapolated from 3 months manifest data
Delaware	300,000	272,000	x	—
Florida	960,000	No data	x	—
Georgia	700,000	38,500,800	Some delisted waste; 99.7% is high volume, aqueous solutions, neutralized on site and discharged to sewers and receiving waters	
Hawaii	30,000	No response		—
Illinois	2,530,000	1,810,000		Manifest data only
Indiana	1,280,000	94,900,000		Includes 92.3 million t of steel industry wastes, pending delisting and currently regulated under NPDES permit
Idaho	80,000	No response		—
Iowa	300,000	No response		—
Kansas	350,000	45,300		Refinery waste, small volume generators
Kentucky	700,000	415,000	x	—
Louisiana	1,250,000	38,800,000	x	—
Maine	130,000	5,290		Mineral spirits, tanning industry waste, small volume generators, substances with LD50
Maryland	590,000	272,100		Waste oil, PCBs, fly ash, and other unspecified waste
Massachusetts	820,000	172,000		Waste oil
Michigan	1,990,000	408,000		Extrapolated from manifest data; 280 compounds (including waste oil) not on EPA list
Minnesota	360,000	181,000		PCBs, crank case oil
Mississippi	340,000	1,810,000	x	—
Missouri	910,000	658,930		Waste oil
Montana	50,000	91,200	x	—
Nebraska	120,000	0.5%		Special waste including infectious waste
Nevada	50,000	No response		—

Table 5 (continued)
HAZARDOUS WASTE GENERATION ESTIMATES BY EPA AND THE STATES[1]

	Quantity (t)		Universe	
State	EPA estimate	State estimate	Same as EPA	State additions to EPA hazardous waste definition
New Hampshire	100,000	9,980		Imported PCBs waste oil
New Jersey	3,120,000	855,000		Manifest data only; waste oil, PCBs, some delisted waste, and other unspecified compounds
New Mexico	60,000	No data		Small volume generators, PCBs, waste oil
New York	2,320,000	1,270,000		PCBs
North Carolina	1,330,000	No response		—
North Dakota	30,000	125,000	x	—
Ohio	2,570,000	3,260,000		Solid waste on a case-by-case basis
Oklahoma	230,000	3,570,000		PCBs
Oregon	200,000	19,100		PCBs and other unspecified compounds
Pennsylvania	2,550,000	3,628,000		Other unspecified compounds
Rhode Island	190,000	1,600		A generally broader definition which includes waste oil, low-level radioactive
South Carolina	1,140,000	1,587,000		Waste oil, paint waste, unstabilized sewerage sludge
South Dakota	10,000	1,590		Waste oil
Tennessee	1,820,000	4,300,000	x	—
Texas	3,010,000	29,146,960		Generally different definition which includes sludge, fly and bottom ash, water soluble oils, boiler sludges, PCBs, and other solid waste
Utah	110,000	558,000	x	—
Vermont	30,000	9,070		Waste oils, infectious waste, PCBs, industrial laundries, some waste delisted by EPA, and other unspecified compounds
Virginia	1,220,000	181,000		PCBs
Washington	380,000	616,000		PCBs
West Virginia	790,000	No response		—
Wisconsin	630,000	81,600		—
Wyoming	40,000	No response		—
Guam	NA	1,450	x	—
Puerto Rico	560,000	417,000	x	—
North Mariana Island	NA	No response		—
American Samoa	NA	0		—
District of Columbia	140,000	No data	x	—
Virgin Islands	NA	No response		—

Table 5 (continued)
HAZARDOUS WASTE GENERATION ESTIMATES BY EPA AND THE STATES[1]

	Quantity (t)		Universe	
State	EPA estimate	State estimate	Same as EPA	State additions to EPA hazardous waste definition
Total	41,200,000 (excluding 2 territories)	250,000,000 (excluding 10 states, 3 territories)		

Note: NA = not applicable.

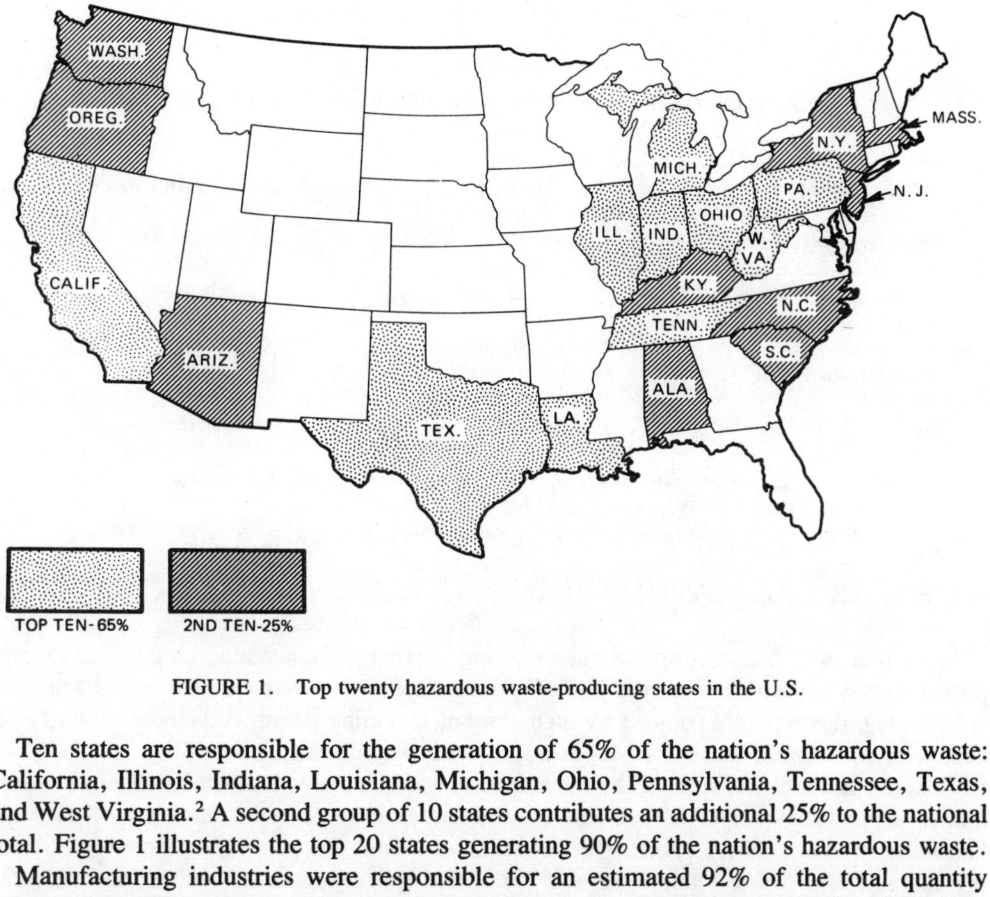

FIGURE 1. Top twenty hazardous waste-producing states in the U.S.

Ten states are responsible for the generation of 65% of the nation's hazardous waste: California, Illinois, Indiana, Louisiana, Michigan, Ohio, Pennsylvania, Tennessee, Texas, and West Virginia.[2] A second group of 10 states contributes an additional 25% to the national total. Figure 1 illustrates the top 20 states generating 90% of the nation's hazardous waste.

Manufacturing industries were responsible for an estimated 92% of the total quantity generated in 1981, with the chemical industry producing approximately 68% of that manufacturing process waste.[16] Table 6 lists the estimated number of hazardous waste generators within five manufacturing groups, compared to the number of nonmanufacturing hazardous material generators.

According to the same survey, 96% of the hazardous waste generated in 1981 was managed on site.[16] However, only 10% of hazardous industrial waste is estimated to be properly handled (i.e., by controlled incineration, secure landfills, and recovery.)[2]

Table 6
ESTIMATED NUMBER OF HAZARDOUS WASTE GENERATORS[16]

Manufacturing	Number
Fabricated metal products	2,636
Chemicals and allied products	2,443
Electrical equipment	1,515
Other metal-related products	2,222
All other manufacturing	3,208
Nonmanufacturing	2,074
Total generators	14,098

Note: Sampling error estimates were ± 25% or less at the 95% confidence level.

Table 7
HAZARDOUS WASTES CURRENTLY INCINERATED[18]

Waste	Quantity incinerated (t/year)	Total quantity incinerated (%)
Characteristic wastes	4,968,915.7	90.0
Nonspecific source wastes	244,655.0	4.4
Specific source wastes	163,325.9	3.0
Acute hazardous wastes	123,795.3	2.2
Toxic wastes	13,399.6	0.2
Miscellaneous wastes	5,359.7[a]	0.1
Total	5,519,451.2	100.0[b]

[a] Ammunition and related waste and trash combined with small quantities of hazardous wastes are included in the miscellaneous waste category.
[b] Percentages do not sum to exactly 100% because of rounding.

VI. HAZARDOUS WASTE STREAMS CURRENTLY BEING INCINERATED

More than 413 hazardous waste streams are currently incinerated, accounting for approximately 5.5 million annually.[17] Table 7 summarizes the respective quantities of hazardous waste categories reported handled by incineration, including the miscellaneous waste fired. Characteristic wastes, that is, wastes exhibiting either ignitability, corrositivy, reactivity, or EP toxicity account for 90% of the hazardous wastes currently being incinerated. Little is known about their composition as they are indicated in the survey data only by their EPA hazardous waste numbers. Corrosive and reactive wastes comprise approximately 86% of the characteristic waste total, 3.5% fall under the ignitable criteria, and EP toxic wastes account for only about 0.13% of the total.[18]

Listed wastes from nonspecific sources compose 4.4% of the total hazardous waste incinerated annually. Slightly more than 93% of this waste is composed of nonhalogenated solvents, while a little more than 6% are halogenated solvents. The principal components of nonhalogenated solvent wastes have not been identified, but the principal constituents of halogenated solvent waste are trichloroethylene, tetrachloroethylene, trichloroethane, and methylene chloride.[18]

Listed wastes from specific industrial sources represent 3% of the total quantity of waste handled annually by incineration. Almost 97% of this type waste is generated by the organic

Table 8
NUMBER OF HAZARDOUS WASTE (HW) INCINERATORS REPORTING MAJOR WASTE QUANTITIES BURNED[19]

EPA waste number	Description	Number of incinerators	Quantity burned (short tons/year)
D001	Ignitables (high Btu, high HW)[c]	60	39,578
D001	Ignitables (low Btu, low HW)[c]	17	140,015
D001	Ignitables (high Btu, low HW)[c]	3	5,870
D001	Ignitables (waste light oils)	2	1,610
D001	Ignitables (unspecified)	5	21,190
D002,3	Corrosives and reactives	30	191,895
F001,2	Spent halogenated solvents	11	17,945
F003	Spent nonhalogenated solvents	18	233,120
F005	Spent nonhalogenated solvents	21	18,253
K011	Acrylonitrile production bottoms	3	120,000
K016—20	Heavy ends, Cl chemical production	10	38,165
K049	Slop oil solids, petroleum ref.	1	5,000
P063	Discarded hydrocyanic acid	5	135,325
U220	Discarded toluene	3	10,801

Note: Information was obtained from an estimated 81% of the HWI facility population. HW = hazardous waste; HWI = HW incinerator; High Btu = 6,000 Btu per pound; High HW = major portion of waste is hazardous (e.g., organic liquids); Low HW = major portion of waste is nonhazardous (e.g., contaminated water).

chemicals industry, with 3% contributed by petroleum refining. Combined, these two industries produce all of the currently incinerated waste in this category.[18]

Combined, acute hazardous and toxic materials account for 2.4% of the total waste incinerated. The major component of nearly all the acute hazardous waste streams is hydrocyanic acid, a potential carcinogen. Toluene, a noncarcinogenic compound, is the constituent found most often in the toxic wastes.[18]

Table 8 exhibits the major types and quantities of waste burned within each classification.

VII. CONSTITUENTS OF INCINERATED HAZARDOUS WASTE STREAMS

A hazardous waste stream can be a complex mixture of many hazardous chemical components, as well as a mixture of both hazardous and nonhazardous constituents. In an EPA survey, a total of 237 different constituents have been identified as present in one or more

Table 9
TWENTY MOST PREVALENT CONSTITUENTS OF HAZARDOUS WASTE STREAMS CURRENTLY INCINERATED[17]

Constituent	Amount incinerated (t/y)
Methanol	133,168
Acetonitrile	58,646
Toluene	56,620
Ethanol	55,090
Amyl acetate	54,926
Acetone	51,535
Xylene	49,453
Methyl ethyl ketone	42,520
Adipic acid	36,135
Ethyl acetate	32,576
1,1,2,2-Tetrachloroethane	29,660
Dichloromethane	29,500
Chlorobenzene	25,871
1,1,1-Tricloroethane	25,789
Cresols	23,312
Dichlorobenzene, NOS	22,954
Trichloroethane	20,436
Cyclohexanol	19,170
Aldicarb	16,850
Phthalic anhydride	15,749

of the 413 hazardous waste streams currently being incinerated.[17] Table 9 lists the 20 most prevalent chemical components of currently incinerated hazardous waste streams.

An important point to remember is that hazardous waste streams and their chemical constituents should be regarded as potential hazards. ''If properly managed, [a hazardous waste] should not cause any actual, significant adverse impact to public health or the environment''.[20] Theoretically then, properly incinerated hazardous waste would not present a risk to human health or the environment.

REFERENCES

1. **Hirschhorn, J. S., Pirages, S., Larsen, K., Goodman, I., Davidson, W. F., Curran, L., Parkinson, R., Sladek, F., and Turnbull, M.,** Technologies and Management Strategies for Hazardous Waste Control, Office of Technology Assessment, Congress of the United States, Washington, D.C., 1983.
2. **Fawcett, H. H.,** *Hazardous and Toxic Materials: Safe Handling and Disposal,* John Wiley & Sons/ Interscience, New York, 1984.
3. Code of Federal Regulations, 40 CFR, 261.10, 1985.
4. **Watson, T., Hall, R. M., Davidson, J. J., and Case, D. R.,** *The Hazardous Waste Handbook,* 4th ed., Government Institutes, Washington, D.C., 1982.
5. Code of Federal Regulations 40, Parts 190 to 399, Revised as of July 1, 1985, Office of the Federal Register, U.S. Government Printing Office, Washington, D.C., 1984.
6. Code of Federal Regulations, 40 CFR, 261.31, 1985.
7. Code of Federal Regulations, 40 CFR, 261.32, 1985.
8. Code of Federal Regulations, 40 CFR, 261.33, 1985.
9. **Lehman, J. P.,** The United States hazardous waste regulatory program, in *Toxic and Hazardous Waste Disposal,* Pojasek, R., Ed., Ann Arbor Science, Ann Arbor, MI, 1980.

10. **Metry, A. A. and Cross, F. L.,** *The Handbook of Hazardous Waste Management,* Technomic Publishing, Westport, CT, 1980.
11. Code of Federal Regulations, 40 CFR, 261.21, 1985.
12. *Webster's Ninth New Collegiate Dictionary,* Merriam-Webster, Springfield, MA, 1985.
13. Code of Federal Regulations, 40 CFR, 261.22, 1985.
14. Code of Federal Regulations, 40 CFR, 261.23, 1985.
15. Code of Federal Regulations, 40 CFR, 261.24, 1985.
16. **Dietz, S. M., Emmet, M., Degaetano, R., Tuttle, D., and Vincent, C.,** National Survey of Hazardous Waste Generators and Treatment, Storage, and Disposal Facilities Regulated Under RCRA in 1981, EPA 530/SW-84-005, Office of Solid Wastes, U.S. Environmental Protection Agency, Washington, D.C., April, 1984.
17. MITRE Corporation, Composition of Hazardous Waste Streams Currently Incinerated, U.S. Environmental Protection Agency, Washington, D.C., April, 1983.
18. Industrial Economics, Inc., Draft Interim Report on Hazardous Waste Incineration Risk Analysis, Office of Solid Waste, U.S. Environmental Protection Agency, Washington, D.C., August, 1982.
19. **Keitz, E., Vogel, G., Holberger, R., and Boberschmidt, L.,** A Profile of Existing Hazardous Waste Incineration Facilities and Manufacturers in the United States, EPA-600/s-84-052, Industrial Environmental Research Laboratory, Office of Research and Development, U.S. Environmental Protection Agency, Cincinnati, OH, February, 1984.
20. **Long, F. A. and Schweitzer, G. E., Eds.,** *Risk Assessment at Hazardous Waste Sites,* ACS Symp. Ser., American Chemical Society, Washington, D.C., 1982.

Chapter 2

INCINERATION TECHNOLOGY

I. INTRODUCTION

Incineration is "an engineered process using controlled flame combustion to thermally degrade waste materials".[1] Proper combustion of a hazardous waste stream will convert the fired waste into nonhazardous byproducts. Several advantages are manifest in the controlled burning process:[2]

1. Incineration is readily applicable to wastes composed of solvents (liquids capable of dissolving and/or dispersing other substances) and other organics.
2. There is little risk of groundwater contamination, which is a disadvantage of land disposal methods.
3. Incinerators can be made to meet air pollution control regulations.
4. The residue (ash) from incineration takes up less space when disposed of in landfills, because the burning process greatly reduces the volume of many wastes.

II. THE BASIC INCINERATION PROCESS

Although individual design, operation, waste applicability, and performance vary among facilities, the basic elements, processes, and goals of hazardous waste incineration are the same. The following sections will describe fundamental incineration technology.

A. Stages in Waste Processing

The typical "cradle to grave" incineration process for a hazardous waste involves the following stages (see Figure 1).

Transportation — The incinerator facility receives incoming waste from liquid-carrying tank trucks, vehicles loaded with drums, or trains or barges transporting containers of waste.

Identification — Wastes are analyzed for content, viscosity, and heating values (if testing did not take place previous to material transport).

Storage — Waste is conveyed to temporary storage tanks.

Mixing — Prior to combustion, wastes may be processed in blending tanks to create mixtures that will burn more readily.

Combustion — Individual feed rates, designed to produce optimum combustion, are employed for pumping waste into the incinerator. At this point, a supplementary fuel feed can be used to further combustion.

Emission control — Following burning, aid gases and particulates are removed from the stack effluent by scrubbers to prevent release to the atmosphere. The scrubber water (as well as the ash) residue is then disposed of.

Based on this generalized waste flow description, the essential components of an incineration system can be identified. First, all hazardous waste incinerators are equipped with a waste feed system, which encompasses the storage and blending units as well as the method for moving material into the combustion area (i.e., a pump or conveyor belt). A second necessary component is a supply of combustion air or oxygen. The combustion mechanism, the third incineration component, includes the firing chamber and any combustion monitors employed. Incinerators may also contain an air pollution control system to clean the exhaust gases before they emit through the stack, some form of ash removal, and equipment for water pollution control to treat scrubber waste before it is discharged.[2] The flow diagram in Figure 2 illustrates the relationship of the basic elements of incinerator design.

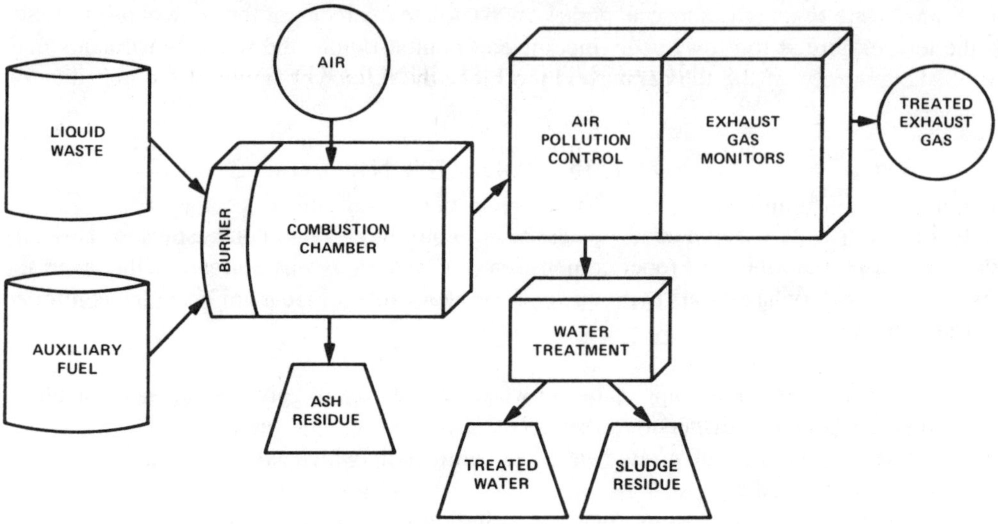

FIGURE 1. Stages of the incineration process for hazardous wastes.

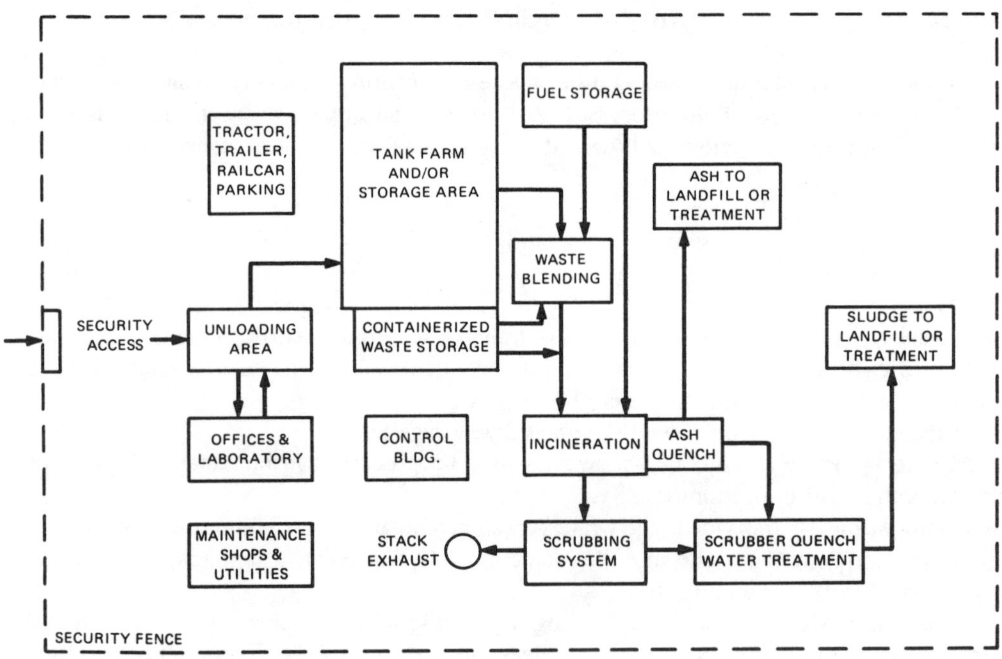

FIGURE 2. Flow diagram illustrating the basic elements of incinerator design.

B. The Three "Ts" of Incineration

Within the combustion mechanism, four design elements are necessary for the proper firing of hazardous waste: oxygen (typically introduced into the system as air) and the three "Ts" of combustion — temperature, time, and turbulence. The oxygen provided is necessary to complete the chemical reaction (oxidation) and functions additionally to control flame temperature. As the primary force that sustains combustion, chamber temperature is the most important of the four design elements. The chamber must be hot enough to activate incoming waste to its ignition point and allow complete oxidation of the compounds present. If the temperature is too low, poor (incomplete) combustion of the waste constituents may result. Conversely, if the temperature is too high, the refractory lining of the unit may be damaged.

Residence time is built into the incinerator through space in the combustion chamber. The volume must be large enough to provide time for (1) adequate exposure of the waste material to high temperatures and (2) complete burning of any solid wastes and volatile gases. A series of baffles (changes in direction) are designed into the combustion chamber to create turbulence during firing. Turbulence mixes the waste material thoroughly with the oxygen molecules, as exposure to the oxygen is necessary to complete the oxidation reaction.[3]

C. Products of Incineration

Carbon dioxide and water are the primary, but not the only, products of the combustion process. Incinerator performance and the characteristics of the waste fired are the major factors which determine the production of potential pollutant substances.[4] The combustion of chlorinated waste, for example, is known to produce hydrogen chloride, a highly corrosive compound. Effluent oxides of metals, sulfur, and nitrogen result from incineration of hazardous wastes which contain those elements originally.

Although little is known about their formation, products of incomplete combustion (PICs) result in small amounts from all combustion sources. PIC compounds, which potentially include highly toxic dioxins and furans, may be very similar to or completely different from the composition of the original chemicals burned. Currently, PIC emissions are not federally regulated, but control proposals have been made.[3] Carbon monoxide is another effluent result of poor combustion, and its presence in the stack gas is useful to measure problems in combustion element functioning (i.e., temperature and turbulence). (See Chapter 7 for a more thorough treatment of PICs.)

D. Pollution Control Devices

The primary function of air pollution control devices (APCDs) is to insure compliance with particulate and acid gas (HCl) emission standards. The installation of APCDs depends upon such factors as incinerator design and waste properties, in addition to applicable federal, state, and local emission laws.[5] High combustion temperatures and long gaseous residence times are two incinerator design elements that would increase the likelihood of having an APCD at a given facility.

Types of pollution control devices employed at incineration facilities include afterburners, scrubbers, and electrostatic precipitators.[3] An afterburner, which is a simple, high temperature combustion chamber adjacent to the primary chamber, aids in the removal of pollutants emitted from the first burner due to incomplete waste combustion. Continued combustion can decrease contaminants in the stack, decrease the pollutant streams to other control devices, and reduce the amount of residue.[4]

Scrubbers operate to remove pollutants from the gas effluent by a liquid spray method. Four common scrubber types are the gas-atomized spray (venturi) scrubber, packed bed scrubber, spray tower, and plate scrubber. An electrostatic precipitator (ESP) is a control method which electrically charges and separates pollutant particles from the stack gas stream. A variation of the ESP is the wet ESP, which uses liquid to wash away the collected particles.[4]

E. Energy Recovery

Twenty-two percent of the operating incinerator facilities responding to a national Environmental Protection Agency (EPA) survey employed a form of waste heat recovery.[5] As incineration temperatures may range up to 1600°C, the flue gas heating value is high enough to make energy recovery beneficial. The most often used waste heat recovery system is the

Table 1
QUANTITIES OF OPERATIONAL HAZARDOUS WASTE INCINERATORS (HWI) REPORTED BY HWI FACILITIES AND MANUFACTURERS[5]

	Reported by HWI facilities contacted	Projection for total population	Reported by manufacturers
Liquid injection	160[a]	213	219
Hearths	56[b]	75	70
Rotary kiln	13[b,c]	17	37
Fluidized bed	4	5	9
Others	31[d]	42	[e]
Type not specified	20	0	[e]
Total operational	284	352	335

[a] Includes fume/liquid units.
[b] Includes units both with and without liquid injection.
[c] Includes two rotary kilns in combination units.
[d] Includes three combination units not having a rotary kiln.
[e] This category was not obtained from manufacturers.

gas-to-water type, although gas-to-air and gas-to-organic fluid types can also be employed at incinerators.[4] Waste energy, most often recovered as steam, can be used to generate electricity, drive machinery, or raise the temperature of combustion air.[3]

III. QUANTITIES OF INCINERATORS

It has been estimated that 350 rotary kiln and liquid injection incinerators are currently used to destroy hazardous waste in the U.S.[6] More realistic estimates are available from manufacturer and incinerator operator survey data,[7] which project a total of 352 for all types of active hazardous waste incinerators (Table 1). The actual number of incinerators verified from the MITRE survey at 219 facilities was 284.[7] Distribution of the 219 facilities arranged by EPA region and state are given in Table 2. Eighty percent of these facilities are privately owned (onsite); commercial and military operations account for the remainder. A typical incineration facility layout is shown in Figures 3 and 4.

IV. TYPES OF INCINERATORS

The principal design features utilized in incinerator classification are combustion chamber type, waste capacity, APCDs, and method of energy recovery.[5] Across types, incinerators burning liquid wastes have a median capacity of 150 gal/h. The average feed rate for solids is 650 lb/h (78 gal/h). EPA survey data report 1800°F as the median combustion temperature, with gaseous residence time averaging slightly less than 2 sec. Table 3 summarizes facility operating data by incinerator type.

A. Liquid Injection

Liquid injection is the most commonly employed thermal destruction technology in the U.S., and it is the only method presently used for incineration at sea.[3] Two major types, horizontally fired and vertically fired, are currently active in hazardous waste destruction. The tangentially fired vortex combustor is a third, less common variation. The combustion mechanisms for all liquid injection incineration systems have simple designs involving very little movement (see Figure 5).

The typical liquid injection firing chamber is a cylinder lined with refractory brick, which

Table 2
LOCATION OF HAZARDOUS WASTE INCINERATION (HWI) FACILITIES BY STATE AND EPA REGION[5]

EPA region/state	Existing HWI facilities Number	Total (%)
Region I	10	4.6
Connecticut	5	2.3
Maine	1	0.5
Massachusetts	3	1.4
New Hampshire	0	0.0
Rhode Island	1	0.5
Vermont	0	0.0
Region II	22	10.0
New Jersey	6	2.7
New York	12	5.5
Puerto Rico	4	1.8
Virgin Islands	0	0.0
Region III	23	10.5
Delaware	2	0.9
District of Columbia	0	0.0
Maryland	4	1.8
Pennsylvania	9	4.1
Virginia	2	0.9
West Virginia	6	2.7
Region IV	46	21.0
Alabama	7	3.2
Florida	7	3.2
Georgia	6	2.7
Kentucky	4	1.8
Mississippi	1	0.5
North Carolina	10	4.6
South Carolina	4	1.8
Tennessee	7	3.2
Region V	29	13.2
Illinois	9	4.1
Indiana	5	2.3
Michigan	2	0.9
Minnesota	2	0.9
Ohio	6	2.7
Wisconsin	5	2.3
Region VI	62	28.3
Arkansas	6	2.7
Louisiana	21	9.6
New Mexico	1	0.5
Oklahoma	3	1.4
Texas	31	14.2
Region VII	8	3.7
Iowa	3	1.4
Kansas	1	0.5
Missouri	2	0.9
Nebraska	2	0.9
Region VIII	5	2.3
Colorado	4	1.8
Montana	0	0.0

Table 2 (continued)
LOCATION OF HAZARDOUS WASTE INCINERATION (HWI) FACILITIES BY STATE AND EPA REGION[5]

	Existing HWI facilities	
EPA region/state	Number	Total (%)
North Dakota	0	0.0
South Dakota	0	0.0
Utah	1	0.5
Wyoming	0	0.0
Region IX	14	6.4
Arizona	1	0.5
California	10	4.6
Hawaii	2	0.9
Nevada	1	0.5
Region X	0	0.0
Alaska	0	0.0
Idaho	0	0.0
Oregon	0	0.0
Washington	0	0.0
Total	219	100.0

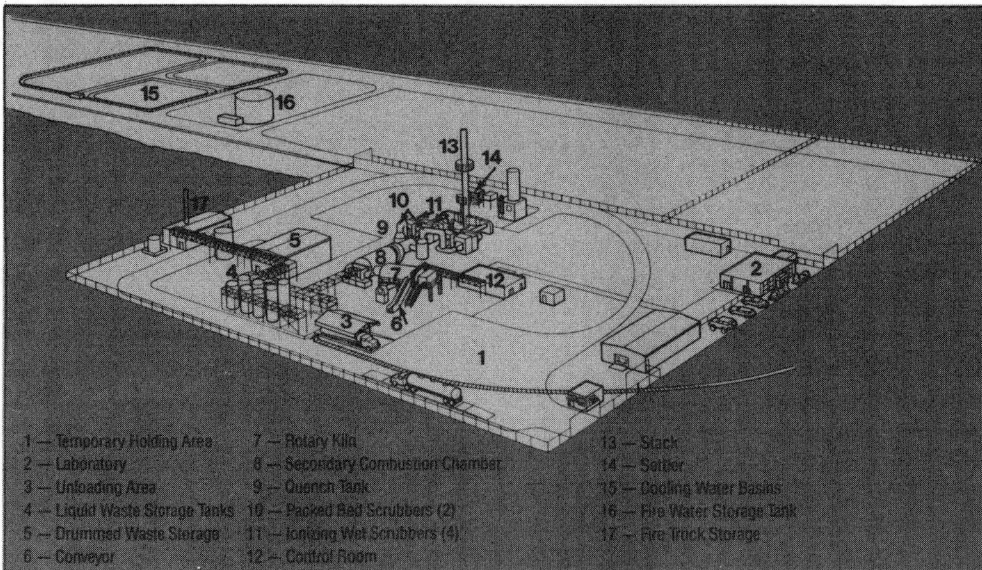

FIGURE 3. Incineration facility layout.

receives hazardous substances through an atomizer nozzle. The atomizer functions to position the flame, maintain continuous ignition, and regulate fuel and air supply, in addition to treating the initial fuel and hazardous waste feed.[3] Because liquid waste must be converted to a gas before it is fired,[4] droplets passing through the nozzle must be small enough to vaporize thoroughly. Small waste droplets are also desirable for completion of the combustion

FIGURE 4. Actual photograph of incineration facility — SCA Chemical Services Facility, Chicago, IL.

Table 3
TYPICAL INCINERATOR OPERATING CONDITIONS AS REPORTED BY MANUFACTURERS[5]

Incinerator type	Combustion zone Temp (°F)	Combustion gas residence time (sec)	Excess air % stoichiometric
Liquid injection	1800—3000	0.3—2.0	120—250
Rotary kiln	1200—2300	2 h (solids)	50—250
Afterburner	2000—2500	1.0—3.0	120—200
Hearth			
Primary chamber	1200—1800	—	30—200
Secondary chamber	1400—2200	1.5—2.5	200—400
Fluidized bed	1400—2000	1.0—5.0	100—150

phases during the brief chamber residence time.[8] Oxygen to enhance the combustion reaction and create turbulence is supplied through a forced draft system.

Waste can be fed into the liquid injection incinerator as rapidly as 5600 l/h. The residence times for combustion chambers range from 0.5 to 2.0 sec, and the temperatures have extreme limits of 700 and 1650°C.[4] Across variable designs, this type of incinerator handles all combustible liquids, encompassing slurries and sludges. Hazardous waste streams that are not well suited for liquid injection firing include noncombustibles such as the heavy metals, streams with high moisture content, inert materials, inorganic salts, and wastes with a high proportion of inorganic components. Although these incinerators characteristically have

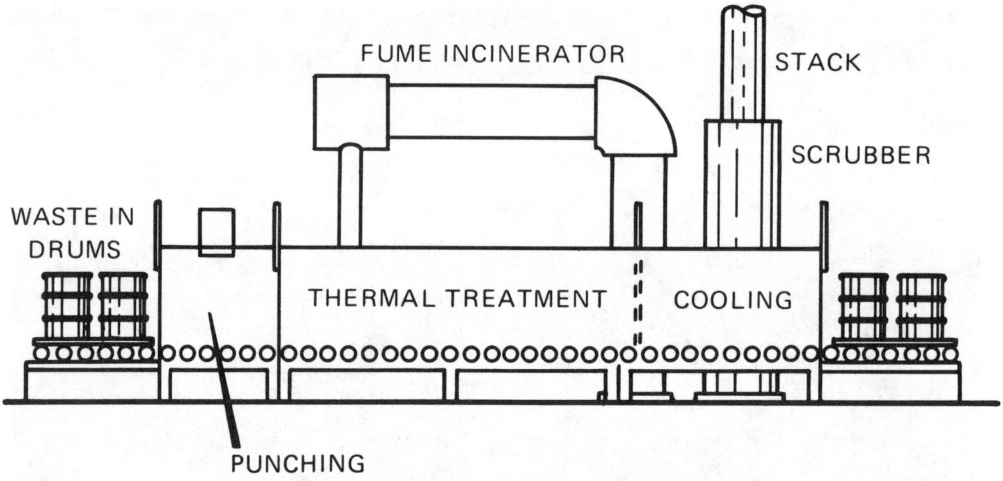

FIGURE 5. Liquid injection incinerator schematic.

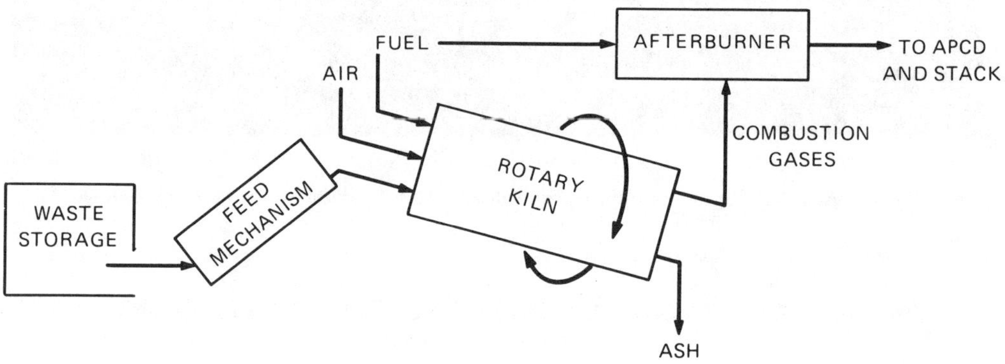

FIGURE 6. Rotary kiln incinerator schematic.

waste-specific operating parameters, many liquid waste mixtures such as pesticides, chemical warfare agents, and motor and industrial oil are destroyed in individually designed liquid injection facilities.[8]

B. Rotary Kilns

Rotary kiln incinerators are highly versatile facilities capable of handling solids, sludges, free liquids, and bases.[8] Two varieties produced in the U.S. are the concurrent type, in which the burner is located at the front of the chamber along with the waste feed, and the countercurrent type, which has a burner at the rear. These burners supply the necessary combustion heat.

The combustion unit is a refractory-lined cylinder which is mounted at an angle and rotates slowly[3] (see Figure 6). Natural gas igniters and gas burners are present to supply the initial refractory heat, maintain flame stability, and provide any necessary supplemental heat. Solid wastes are typically fed into the high end of the kiln, while liquids and gaseous wastes are pumped into the cylinder simultaneously through atomizing nozzles. Solids are vaporized and ignited with air through the chamber rotation. The rotation assures uniform heat transfer and is responsible for depositing the ash at the lower end of the combustion unit.[3] An additional combustion chamber, or afterburner, is often present to assist in more complete combustion (i.e., treating flue gases before they reach the APCD).

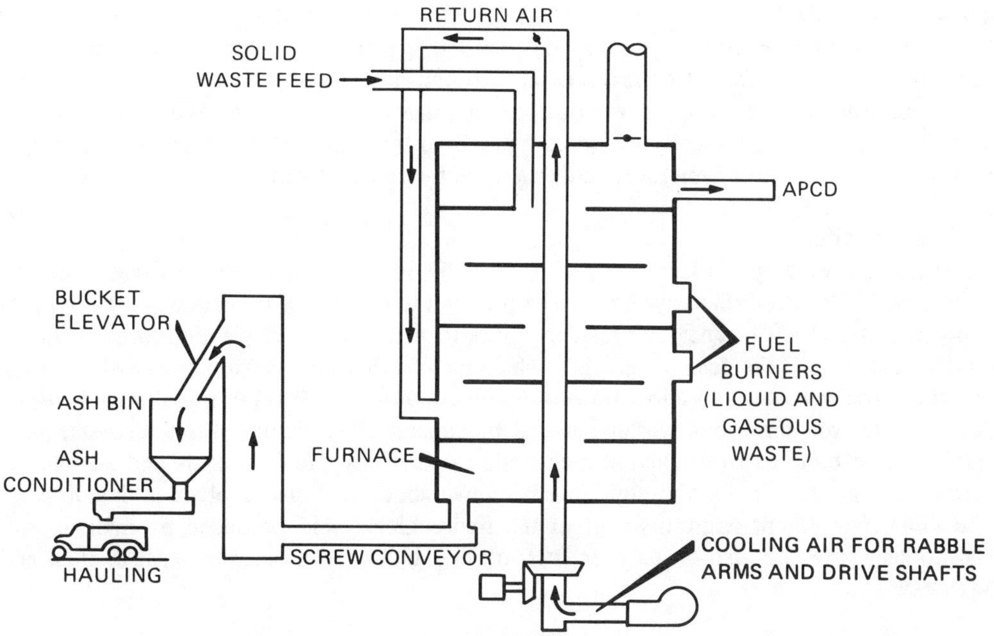

FIGURE 7. Multiple hearth incinerator schematic.

The cylinders have a diameter range of 3 to 20 feet, with a length variation of 8 to 30 feet. Kiln temperatures can be as low as 1500°F or as high as 3000°F. Waste loading capacities vary between 1 and 8 tons/h.[8] Rotary kiln firing is generally applicable to any combustible hazardous waste and notable for destruction of polyvinyl chloride wastes, PCBs, chemical warfare agents, and munitions. Here again, those wastes that are not well suited to this method of burning are inorganics, heavy metals, and wastes composed primarily of moist or inert materials.

C. Multiple Hearths

A multiple hearth furnace consists of a central rotating shaft that has three zones of operation (see Figure 7). Into each zone of operation are built cross-sectional floors (hearths) with drop holes. Depending on the zone, temperatures range from 200 to 1000°C for waste treatment. The first zone receives sludge and/or granulated solid waste through the steel shell roof from a conveyor. Waste feed in the upper hearths is dried before it falls through drop holes to the next level. The central hearths compose the burning zone, where agitation exposes the waste surface to hot gases as it moves across each hearth. In the cooling zone, the waste is treated by the remaining hearths until it is discharged as ash at the bottom of the series.[4]

Multiple hearth incineration is most successful for sludge disposal, although auxiliary nozzles can inject liquids and gases. Solids, generally pretreated, can also be fired by this method. Sludges treated by multiple hearths (i.e., municipal sewage) are generally burned onsite at industrial plants.[8] Some problems preventing more widespread use of this method include evidence of cold spots and short-circuited residence times.

D. Fluidized Bed

Thermal destruction of hazardous waste by the fluidized bed method is a relatively simple process in which an inert granular bed (usually sand) transfers combustion heat to injected material.[8] Bed heat transfer is quick and thorough and also provides for the oxidation of gases above the bed. The fired waste then returns the energy to the granules. The fluidized bed operates at temperatures between 450 and 850°C and has a depth of 2 m. Turbulence is maintained by the forced entrance of air through a perforated distributor plate. Liquid hazardous wastes reside in the combustor approximately 12 to 14 sec. The wastes handled by a fluidized bed are typically the same types fired by rotary kilns; the granular material, however, may have heat limitations and thus limit the range of suitable waste feed streams.[4]

E. Cement Kilns

Cement kilns are specialized types of rotary kilns in which liquid organic wastes can be cofired with the kiln flame base fuel.[4] The production of cement necessitates high temperatures (2600 to 3000°F) and long residence times for the chemical reaction to occur in the product material. These design elements, which include thorough mixing, are ideal for firing difficult-to-burn wastes and allow for processing 1000 to 25,000 gal of liquid organic waste per day. Suitable hazardous wastes handled by cement kilns include waste solvents, paint residues, and bottoms from solvent reclamation operations. Heavy metals and ash are incorporated into the cement although little is known about the risks involved in this process. The kiln environment neutralizes all of the hydrochloric acid produced by incineration. Hazardous waste can also be used as fuel in other industrial processes, such as lime and aggregate kilns.[3]

F. Boilers

Coincineration is the firing of a hazardous waste stream mixed with another substance, such as refuse or sludge. It is used to increase production of Btus for complete combustion of low heat content waste. Coincineration can occur in any existing facility, but the most prevalent form in the U.S. is the ignition of wastes with primary fuels in boilers.[8]

Power boilers, which are similar in design to liquid injection incinerators, convert the heat of combustion into steam. They are estimated to destroy twice the amount of hazardous waste than all other incinerators combined (5.5 million tons in 1981). Coupled with the lack of RCRA regulations, the energy recovery factor has been instrumental in the widespread use of boilers by the organic chemical industry. Four hundred thousand boilers, including industrial, commercial, residential, and off-site power units, currently operate nationwide.[3] Information regarding effective hazardous waste destruction is lacking, however, due to the nonregulated status of coincineration.

Two factors that may inhibit complete combustion in boilers and increase dangerous emissions are (1) the walls are cooled in order to make use of the energy from the primary product gas, which may result in combustion products passing through a cool layer and (2) some internal parts of boilers are metallic rather than refractory, making them vulnerable to corrosion by halogenated organic waste firing.

G. Ocean Incineration

Incineration at sea is currently an adaptation of liquid injection technology to ships, without scrubbers for the stack exhaust. Scrubbers are not required due to the water and sea sir functioning as a buffer.[8] The removal of APCDs enhances the combustion capability of these incinerators. The surface ships, Vulcanus I and II, presently the only completed ocean-based incineration facilities, are equipped with incinerators mounted on the stern, each with burning capacities of 7 $^1/_2$ t/h (1650 gal/h). Temperatures range from 1166 to 1600°C during combustion; residence times for hazardous waste average 1 sec. Combustion air is provided by a large fan. As only liquid wastes are burned, afterburners are not utilized and particulate emissions control is not required.[3] Heat recovery mechanisms are not employed in this technology as the energy would not be used at the location, and the hydrogen chloride emissions could cause damage to recovery equipment.

H. Pyrolysis/Starved Air Combustion

Pyrolysis denotes thermal decomposition of a hazardous waste compound in the absence of oxygen. Starved air combustion uses oxygen, but less than the minimum theoretical amount required for complete combustion. A pyrolyzing chamber and a fume incinerator are the components of a pyrolysis facility,[8] although a pyrolyzer can be an inclusion in another type of incineration unit such as a rotary kiln.[4] Pyrolysis destroys wastes between temperatures of 1000 and 1700°F. This method serves to convert both pure carbon and carbon compound waste into usable substances. The federal government has used pyrolysis in the destruction of chemical warfare agents and sludges containing kepone. Industrial use of this process includes disposal of rubber scrap, pharmaceutical biosludge, and organic chlorine sludge. All phases of compounds containing carbon, hydrogen, and/or oxygen are applicable to the oxygen-deficient method, while unsuitable wastes are those containing nitrogen, sulfur, sodium, silicon, phosphorus, fluorine, bromine, chlorine, and/or iodine.

I. Alternative Thermal Destruction Technologies

In addition to conventional incineration processes, several alternative thermal destruction methods are either currently available or projected to be commercially available for hazardous waste management within the next 5 years. These emerging alternatives include high temperature electric reactors, molten salt reactors, plasma arc reactors, wet air oxidation, supercritical water, and molten glass incineration.

1. High Temperature Electric Reactors

Hazardous waste incineration is accomplished in high temperature electric reactors through radiation of a cylindrical core.[8] The reactor walls are heated by implanted carbon electrodes, and the core is surrounded by a nitrogen-formed inert gaseous layer. Rapid heat transfer to the waste occurs when the core reradiates the electric energy. The formation of PICs is rare due to the quick heat transfer and extremely high temperatures. The organic wastes burned, such as PCB-contaminated soils, are heated to temperatures ranging from 2000 to 2400°C. The reactors can burn hazardous solids at a rate of 75 to 125 lb/min.[3]

2. Molten Salt Reactors

A molten salt reactor oxidizes liquid, solid, or gaseous hazardous wastes that are injected into a pool of molten sodium or calcium carbonate. This process, in which bed temperatures average 900°C, is applicable to highly toxic wastes and chemical warfare agents. In the pool, organic constituents are broken down into carbon dioxide and water, while the sodium carbonate handles combustion byproducts, neutralizes HCl, and scrubs the off-gases.[8] The inorganic salts (combustion byproducts) retained in the pool eventually build up and must be removed. In pilot scale tests, extremely high levels of hazardous constituent destruction have been achieved with molten salt reactors.

3. Plasma Arc Reactors

In this thermal destruction method, electrical energy is transformed into a plasma. The plasma, a gas composed of both charged and neutral particles which can conduct electricity, can reach temperatures of 50,000°F or more.[8] Waste materials receive heat from the decay of the activated plasma components. These exposed wastes are converted to gaseous form, ionized, and are destroyed almost immediately. This rapid destruction at such high temperatures omits the reaction progression which can allow the formation of PICs. Highly toxic liquid wastes can be burned at a rate of 600 lb/h in plasma-arc reactors. Currently, there is no environmental data available with regard to plasma-arc destruction performance.[3]

4. Wet Air Oxidation

Aqueous organic waste streams which are too dilute for conventional incineration and too toxic for biological treatment are applicable to wet air oxidation (WAO). A reactor is used to contain this process in which thermal destruction of organic contaminants occurs in the liquid phase.[8] The waste stream undergoes preheating via heat exchange, increasing the temperature enough to support the oxidation reaction with the injected liquid air. During oxidation the temperature increases, and the heat of combustion is recovered for use within the process of heat exchange. Water (the primary component of aqueous waste streams) modifies the reaction temperatures and serves as a heat transfer agent. The oxidation temperatures are comparatively low with regard to most other thermal destruction technologies, ranging from 350 to 650°F.[3] Excessive evaporation of the liquids is prevented by maintaining pressure between approximately 300 and 3000 psi. Information regarding environmental impact of WAO is limited, but EPA evaluation of the process is currently underway.

5. Supercritical Water

At temperatures and pressures above the critical point (greater than 374°C and 218 atmospheres), water takes on properties that enable it to destroy aqueous hazardous waste stream components.[3] Supercritical water is a solvent for organic compounds and is capable of breaking down large organic molecules.[8] Oxidation of organics occurs rapidly with this method because oxygen mixes thoroughly with supercritical water. Inorganics are easily separated from applicable waste streams due to their virtual insolubility in this phase of water. The resulting effluent streams from a supercritical process are free of both types of contaminants. This technology may evolve as a preferred destruction method for such hazards as DDT and PCBs, as well as other aqueous wastes with high levels of inorganics and toxic organic components. Further testing is needed to expand the limited environmental data.[3]

6. Molten Glass Incineration

Any combustible waste can be fired in a molten glass furnace, but no hazardous waste incineration data for this method is currently available.[3] A molten glass furnace, which is used in the glass-manufacturing industry, consists of a combustion chamber with a pool of molten glass covering the bottom. Temperatures in the glass pool above 2300°F (1260°C) are supported by electrodes. Combustible wastes enter the combustion chamber and are oxidized above the molten glass; inorganic substances and ash are incorporated into the pool. Ceramic filters control the exhaust gases and are ejected into the molten glass when new replacements are needed. With the exception of molten glass incineration, Table 4 compares thermal destruction technologies for hazardous waste and lists the utilization status for each in the U.S.

V. PRINCIPAL ORGANIC HAZARDOUS CONSTITUENTS

Theoretically, to determine the effectiveness of an incinerator in destroying hazardous waste components, one would like to know the percentage of the waste stream that is destroyed. Since it is difficult to measure for all chemical compounds present in a waste stream, representative chemicals are chosen from waste streams for testing. These compounds, known as principal organic hazardous constituents (POHCs), are chosen to represent the complex matrix of the fired waste. The POHC system, developed by the RCRA program, employs an EPA ranking of incinerability for the compounds.[3] The ranking is determined by a compound's heat of combustion, which provides an indication of how difficult a compound is to burn. The heat of combustion is a theoretical calculation of energy released when waste molecules are combusted. The assumptions of the POHC selection system are that (1) compounds with lower heats of combustion are more difficult to burn than those with high heats of combustion and (2) if a difficult to incinerate compound is destroyed effectively, then all compounds with higher heats of combustion than that compound will

Table 4
COMPARISON OF THERMAL TREATMENT TECHNOLOGIES[6]

Advantages of design features	Disadvantages of design features	Status for hazardous waste treatment
	Currently available incinerator designs	
Liquid injection incineration: can be designed to burn a wide range of pumpable waste and is often used in conjunction with other incinerator systems as a secondary afterburner for combustion of volatilized constituents; hot refractory minimizes cool boundary layer at walls; HCl recovery is possible	Limited to destruction of pumpable waste viscosity of less than 10,000 SSI); is usually designed to burn specific waste streams; smaller units sometimes have problems with clogging of injection nozzle	Estimated that 219 liquid injection incinerators are in service making this the most widely used incinerator design
Rotary kilns: can accommodate a great variety of waste feeds (solids, sludges, liquids, and some bulk waste contained in fiber drums); rotation of the combustion chamber enhances mixing of waste by exposing fresh surfaces for oxidation	Rotary kilns are expensive; economy of scale means regional locations, thus waste must be hauled increasing spill risks	Estimated that 42 rotary kilns are in service under interim status; rotary kiln design is often the centerpiece of integrated commercial treatment facilities; first noninterim RCRA permit for a rotary kiln incinerator (IT Corp.) is currently under review
Cement kilns: attractive for destruction of harder-to-burn waste due to very high resistance times, good mixing, and high temperatures; an alkaline environment neutralizes chlorine	Burning of chlorinated waste is limited by operating requirements and appears to increase particulate generation; could require retrofitting of pollution control equipment and of instrumentation for monitoring to bring existing facilities to comparable level; ash may be a hazardous residual	Cement kilns are currently in use for waste destruction, but the exact number is unknown; national kiln capacity is estimated at 41.5 million/year; at present, mostly nonhalogenated solvents are burned
Boilers (usually a liquid injection design): energy value recovery and fuel conservation; availability on sites of waste generators reduces spill risks during hauling	Cool gas layer at walls results from heat removal; this constrains design to high efficiency combustion within the flame zone; nozzle maintenance and waste feed stability can be critical; where HCl is recovered, high temperatures must be avoided (high temperatures are good for DRE.); metal parts corrode where halogenated wastes are burned	Boilers are currently used for waste disposal; the number of boiler facilities is unknown; the quantity of wastes combusted has been roughly estimated at between 17.3 to 20 million t/year
	Applications of currently available designs	
Multiple hearth: passage of waste onto progressively hotter hearths for long residence times for sludges; good fuel efficiency; ability to handle wide variety of sludges	Tiered hearths usually have some relatively cold spots which inhibit even and complete combustion; opportunity for some gas to short circuit and escape without adequate residence time; not suitable for waste streams which	Technology is available, widely used for coal and municipal waste combustion

Table 4 (continued)
COMPARISON OF THERMAL TREATMENT TECHNOLOGIES[6]

Advantages of design features	Disadvantages of design features	Status for hazardous waste treatment
Applications of currently available designs (continued)		
	produce fusible ash when combusted; units have high maintenance requirements due to moving parts in high temperature zone	
Fluidized-bed incinerators: turbulence of bed enhances uniform heat transfer and combustion of waste; mass of bed is large relative to the mass of injected waste	Limited capacity in service; large economy of scale	Estimated that nine fluidized-bed incinerators are in service; catalytic bed may be developed
At-sea incineration: minimum scrubbing of exhaust gases; limited burns of organochlorine and PCB are required by regulations on assumptions that were conducted at sea in mid-1970; PCB ocean water provides sufficient test burns conducted by Chemical Waste neutralization and dilution; this could provide economic advantages over land-based incineration methods; also, incineration occurs away from human populations; shipboard incinerators have greater combustion rates, e.g., 10 t/h	Shipboard (usually liquid injection incinerator): not suitable for wastes that are shock sensitive, capable of spontaneous combustion, or chemically or thermally unstable due to the extra handling and hazard of shipboard environment; potential for accidental release of waste held in storage (capacities vary from between 4000 to 8000t)	Movement seems to be away from at-sea incineration at this time; no new ships under construction.
At-sea incineration: same as above, except relative stability of platform reduces some of the complexity in designing to accommodate rolling motion to the ship	Oil drilling platform-based: requires development of storage facilities; potential for accidental release of waste held in storage	Proposal for platform incinerator reviewed by EPA; shown to be conditionally feasible, but has not been pursued.
Pyrolysis		
Air pollution control needs minimum: air-starved combustion avoids volatilization of any inorganic compounds; these and heavy metals go into soluble solid char; have potentially high capacity	Greater potential for PIC formation; for some wastes produce a tar which is hard to dispose of; potentially high fuel maintenance cost; waste-specific designs only	Are commercially available but in limited use
Emerging thermal treatment technologies		
Molten salts act as catalysts and efficient heat transfer medium: self-sustaining for some wastes; reduce energy use and reduce maintenance costs; units are compact, potentially portable; minimal air pollution control needs,	Commercial-scale applications face potential problems with regeneration or disposal of ash-contaminated salt; are not suitable for high ash wastes; chamber corrosion can be a problem; avoiding reaction vessel corro-	Technology has been successful at pilot plant scale and is commercially available

Table 4 (continued) COMPARISON OF THERMAL TREATMENT TECHNOLOGIES[6]

Advantages of design features	Disadvantages of design features	Status for hazardous waste treatment
Emerging thermal treatment technologies (continued)		
some combustion products, e.g., ash and acidic gases are retained in the melt	sion may imply tradeoff with DRE	
High temperature fluid wall: waste is efficiently destroyed as it passes through cylinder and is exposed to radiant heat temperatures of about 4000°F; cylinder is electrically heated, heat is transferred to waste through inert gas blanket, which protects cylinder wall; mobile units possible	To date, core diameters (3, 6, and 12 in.) and cylinder length (72 in.) limit throughout capacity; scale-up may be difficult due to thermal stress on core; costs potentially high for electrical heating	Other applications tested, e.g., coal gasification, pyrolysis of metal-bearing refuse and hexachlorobenzene (test burns on toxic gases in December 1982)
Plasma arc: very high energy radiation (at 50,000°F) breaks chemical bonds directly without series of chemical reactions; extreme DREs possible with no or little chance of PICs; simple operation, very low energy costs, mobile units planned.	Limited throughout; high use of NaOH for scrubbers	Limited U.S. testing, no commercialization underway.
Wet oxidation: applicable to aqueous waste too dilute for incineration and too toxic for biological treatment; lower temperatures required, and energy released by some wastes can produce self-sustaining reaction; no air emissions	Is not applicable to highly chlorinated organics; some wastes need further treatment	Commercially used as pretreatment to biological wastewater treatment plant; bench-scale studies performed with catalyst for nonchlorinated organics
Super critical water: applicable to chlorinated aqueous wastes which are too dilute to incinerate; takes advantage of excellent solvent properties of water above critical point for organic compounds; injected oxygen decomposes smaller organic molecules to CO_2 and water; no air emissions	Probable high economy of scale; energy needs may increase on scale-up	Bench-scale success (99.99% DRE) for DDT, PCBs, and hexachlorobenzene

also be destroyed. In general, the POHCs most abundant in a waste stream and presumed the most difficult to destroy are the compounds isolated for destruction measurements.

Due to concern that the heat of combustion rankings and POHC selections are inadequate to determine incinerability for complex chemical mixtures of wastes, the EPA is currently evaluating the system's reliability. However, the use of POHCs is currently supported by the EPA as the best available method for determination of incinerator efficiency.

VI. DESTRUCTION EFFICIENCY (DE) AND DESTRUCTION AND REMOVAL EFFICIENCY (DRE)

Baseline hazardous waste incinerator performance is primarily determined by calculating

(1) destruction efficiency (DE) and (2) destruction and removal efficiency (DRE) for hazardous constituents.[9] DE measures the percentage of an individual constituent destroyed in the combustion chamber. DRE calculations encompass combustion chamber destruction of a hazardous waste component and further removal of that same constituent by air pollution control equipment. DE/DRE is the most important consideration involved in trial burns (see the section on trial burns).

The POHC system described above is employed to select a waste component for DRE measurement. The RCRA has set an incinerator performance standard of 99.99% DRE for POHCs,[10] which is based on the following formula:

$$DRE = \frac{W_{in} - W_{out}}{W_{in}} \times 100$$

where W_{in} = mass feed rate of a constituent in the waste stream feeding the incinerator; and W_{out} = mass emission rate of the same constituent in the exhaust emissions prior to release to the atmosphere.

Based on test run findings at five incinerators, APCD scrubbers, which are utilized in removing hazardous constituents from stack emissions, do not seem to significantly improve DRE measurements.[3] In the test situation, POHCs were generally either below detection limits or present only in small amounts in scrubber residue, which contains all hazardous materials removed from the stack. Compared to the scrubber water, POHCs in stack gas were found in much higher concentrations.

Another measure of incineration effectiveness, in addition to DRE, is combustion efficiency (CE). CE indicates how much of the carbon in the waste is converted into carbon dioxide (CO_2), rather than into carbon monoxide (CO). CO is a result of incomplete combustion of compounds and therefore lends itself to an evaluation of an incinerator's overall capability for high levels of waste destruction. Although high CEs generally correlate with high DE/DRE measurements, no precise relationship between the two has been determined. Presently, CE measurements are not a federal requirement, but individual incinerator operating permits include maximum stack gas levels of CO.[3]

VII. TRIAL BURNS

A hazardous waste incinerator must demonstrate the ability to comply with EPA performance standards before an operating permit is granted. Trial burns are the typical method to determine such incinerator capability. A trial burn is "any attempt to incinerate the waste in question for a limited period . . . designed to establish the conditions at which incineration of waste in a given facility must be carried out to assure protection to public health and the environment".[4] A temporary permit is issued for the duration of a trial burn, which can involve one test run or a testing series. The test burns should not pose an environmental or health risk; thus, provisions must be made for rapid detection of dangerous emission quantities and subsequent immediate shutdown of the facility.

Determination of DRE is the primary objective of a hazardous waste incinerator trial burn. However, the amount of data gathered extends far beyond that which is necessary for DRE calculations. First, potential waste streams treated at the facility should be analyzed for heating value, POHC concentrations, viscosity, and chlorine and ash content. POHC concentrations must also be determined in stack effluent, bottom ash, and scrubber residue. Sampling and monitoring data are collected for calculation of hydrogen chloride (HCl) removal, amount of particulate emissions, identification and sources of fugitive emissions, combustion temperature and gas velocity range, and continuous measure of CO in stack exhaust. Based on these measurements, the incinerator's performance level must compare

Table 5
TECHNICAL PERFORMANCE STANDARDS FOR INCINERATORS[6]

Design and operations conditions	Inspection and monitoring	Closure/post closure
	Incinerators (subpart 0)	
Incinerators must achieve 99.99% DRE for POHCs specified in permit and specific emissions limits for HCl and particulates set in permit; facility must install monitoring equipment and process controls necessary to assure operation within permit limits at all times to control fugitive emissions and to provide automatic waste feed shutoff if operations exceed permit conditions; permit will specify acceptable range in composition and operating limits for each waste feed; facility must burn only waste feeds approved in permit under specified operating conditions based on trial burn or alternative data; exemptions are facilities burning only ignitable, corrosive, or reactive wastes that contain no, or insignificant amounts of, Appendix VII constituents and that will not pose a threat to human health or the environment if incinerated; these facilities must be permitted and comply with RCRA Subpart 0 waste analysis and closure standards and with general CFR Part 264 treatment, storage, and disposal facility (TSDF) standards	Submit trial burn emissions monitoring results for different waste feeds for permit; special trial burn permit for new facilities required; during operation daily visually inspect incinerator and equipment for spills, leaks, fugitive emissions, continuously monitor temperature, feed rate, air flow, stack gas CO, and other indicators of operating conditions and possible malfunctions; at RA request, sample and analyze waste feeds and stack gas emissions to verify compliance with standard; test weekly waste feed cutoff system and alarms unless RA specifies less frequent period	Remove wastes and residues, decontaminate equipment; send hazardous wastes to TSDF

favorably with three federal regulations: (1) DRE of 99.99% for hazardous compounds, (2) less than 1.8 kg/hr of HCl emission or 99% chlorine removal efficiency, and (3) 0.08 grain particulate emission per dry standard cubic foot. Table 5 summarizes federal incinerator standards under the current program.

Following an incinerator's adequate demonstration of the trial burn requirements, an operating permit is issued. The final permit will stipulate case by case operational requirements for five parameters:[10] (1) CO level in stack exhaust (indirect measure of CE), (2) waste feed rate, (3) combustion zone temperature, (4) combustion gas flow rate (reveals residence time), and (5) APCD operations.

REFERENCES

1. **Brunner, C. R.,** *Incineration Systems: Election and Design,* Van Nostrand Reinhold, New York, 1984.
2. **Metry, A. A. and Cross, F. L.,** *The Handbook of Hazardous Waste Management,* Technomic Publishing, Westport, CT, 1980.

3. U.S. Environmental Protection Agency, Assessment of Incineration as a Treatment Method for Liquid Organic Hazardous Waste, Background Report I: Description of Incineration Technology, Washington, D.C., March, 1985.
4. **Bonner, T. A., Cornett, C. L., Desai, B. O., Fullenkamp, J. M., Hughes, T. W., Johnson, M. L., Kennedy, E. D., McCormick, R. J., Peters, J. A., and Zanders, D. L.,** Engineering Handbook for Hazardous Waste Incineration, U.S. Environmental Protection Agency, Washington, D.C., September, 1981.
5. **Keitz, E., Vogel, G., and Holberger, R.,** Profile of Existing Hazardous Waste Incineration Facilities and Manufacturers in the United States, EPA-600/2-94-052, Industrial Environmental Research Laboratory, U.S. Environmental Protection Agency, Cincinnati, OH, 1984.
6. Office of Technology Assessment (OTA), Technologies and Management Strategies for Hazardous Waste Control, OTA-M-196, U.S. Government Printing Office, Washington, D.C., March, 1983.
7. MITRE Corporation, Composition of Hazardous Waste Streams Currently Incinerated, U.S. Environmental Protection Agency, Washington, D.C., April 1983.
8. **Hirschhorn, J. S., Pirages, S., Larsen, K., Goodman, I., Davidson, W. F., Curran, L., Parkinson, R., Sladek, F., and Turnbull, M.,** Technologies and Management Strategies for Hazardous Waste Control, Office of Technology Assessment, Congress of the United States, Washington, D.C., March, 1983.
9. **Trenholm, A., Gorman, P., and Jungclaus, G.,** Performance Evaluation of Full-Scale Hazardous Waste Incinerators, Vols. I-V, EPA-600/2-84-181a-e, Industrial Environmental Research Laboratory, U.S. Environmental Protection Agency, Cincinnati, OH, November, 1984.
10. Code of Federal Regulations, 40, Parts 190 to 399, July 1, 1985, Office of the Federal Register, U.S. Government Printing Office, Washington, D.C., 1985.

Chapter 3

SUMMARY OF TRIAL BURN DATA

I. INTRODUCTION

Fifty-four trial burns measuring destruction efficiencies (DE) for hazardous waste incinerators were conducted or identified by the Environmental Protection Agency (EPA) during the 1970s.[1] The incinerator types utilized in the 54 tests included such technologies as molten salt, fluidized bed, and pyrolysis, in addition to the more conventional liquid injection, rotary kiln, and hearth incinerators. A DE of at least 99.99% for compounds was achieved in 45 of the 54 burns. The present Federally regulated destruction and removal efficiency (DRE) standard of 99.99% was developed from the data gathered in these test burns.

In 1981, EPA decided to conduct test burns at eight additional facilities to expand the data base for incinerator performance as part of an incinerator regulatory impact analysis (RIA). To prepare for these eight trial burns, EPA conducted a test burn at the Cincinnati Metropolitan Sewage District (MSD) incinerator to verify the EPA trial burn protocol. A brief description of these trial burns follows.

II. SITE AND WASTE CHARACTERISTICS

Following preliminary testing of the EPA trial burn protocol at the Cincinnati MSD incinerator, EPA conducted trial burns at the following eight hazardous waste incinerator facilities: Ross Incinerator Services, Inc., Grafton, OH; Upjohn Company, LaPorte, TX; Zapata Industries, Inc., Butner, NC; American Cyanamid, Willow Island, WV; Mitchell Systems, Inc., Spruce Pine, NC; DuPont Ponchartrain Works, LaPlace, LA; and Trade Waste Incineration, Inc., Sauget, IL. The remaining incinerator was designated Plant B.

These facilities were selected to exhibit a broad range of incinerator design, operating parameters, and waste characteristics. Parameters considered important included (1) incinerator type; (2) operating factors such as combustion temperature, residence time, heat input rate, and excess air; (3) performance of emissions control equipment; and (4) waste characteristics such as organic components, moisture content, heating value, and ash and chloride content.

A. Site Descriptions

The Cincinnati MSD incinerator had both a rotary kiln and a cyclone furnace for burning waste and was equipped with air pollution control mechanisms. Of the other trial burn facilities, all were equipped with liquid injection burners; two of the sites also had rotary kiln incinerators; hearth burners were in use at two facilities; and one plant had a gas injection incinerator. Five of the eight had air pollution control devices (APCDs), while three incinerators did not. The five incinerators with APCDs had scrubbers for HCl control, and four of those five sites were equipped to handle particulate (vapor) emissions.

The incinerators tested displayed a wide range in combustion temperature, residence time, heat input, and percent oxygen in the stack. Characteristics of seven of the eight facilities are summarized in Table 1. For complete treatment of site-specific design and operation parameters of the eight facilities, see Treholm et al.[2] Full description of the preliminary trial burn facility in Cincinnati is available in Gorman and Ananth.[3]

B. Waste Descriptions

At the Cincinnati MSD incinerator, only liquid wastes were burned, although the kiln

Table 1
SUMMARY OF KEY PROCESS PARAMETERS MEASURED DURING TRIAL BURNS[2]

Facility	Run no.	Temp (°C)	Residence/time (sec)	Heat input (10^6 kJ/h)	Excess oxygen (% O_2 in stack)
Ross	1	1154	6.2	87	10.4
	2	1146	6.5	60	10.5
	3	1117	6.7	87	10.7
Upjohn	2	1116	5.2	6.2	8.1
	3	1116	5.2	6.3	8.3
	4	1116	5.2	6.1	8.4
Zapata	1	671/854[a]	0.069	1.5/1.9[b]	8.2
	2	721/868	0.067	1.7/2.1	12.0
	3	738/845	0.066	1.0/1.5	11.8
	4	690/905	0.063	0.70/1.0	11.9
American	1	779[c]	0.17	5.0	10.3
Cyanamid	2	671	0.21	5.0	12.4
	3	629	0.24	4.4	No data
	4	679	0.23	4.7	12.7
	5	648	0.21	4.5	13.1
Mitchell	1	1052/1010[d]	2.5	8.3	9.4
	2	1065/1093	2.4	9.0	9.5
	3	1093/1121	2.2	10.5	No data
	4	1079/1079	2.2	6.2	10.8
DuPont	1	807/1000/1450[e]	7.2/0.26[f]	19/17[g]	9.2
	2	750/975/1450	7.0/0.25	17/17	9.6
	3	750/967/1450	7.6/0.28	19/17	10.3
TWI	1	1137/1110[h]	4.7	7.7	12.4
	2	1110/1093	3.5	7.9	13.0
	3	1132/1110	3.5	6.3	13.2
	4	988/966	3.0	7.4	15.6

[a] First value is main chamber temp, second value is secondary chamber temp.
[b] First value represents only waste heat input, second value includes natural gas heat input.
[c] Temp measured in stack near entrance.
[d] First value is upper chamber temp, second value is lower chamber temp.
[e] First value is kiln temp, second temp is afterburner temp, third value is liquid injector temp.
[f] First value is the kiln/afterburner residence time, the second value is the liquid injector residence time.
[g] First value is kiln heat input, second value is liquid injector heat input.
[h] First value is primary chamber temp, second value is secondary chamber temp.

could also fire solid and semisolid wastes. The liquids fired during the trial burn included one pesticide-containing waste and one waste high in chlorine content.

The wastes burned at the remaining eight facilities were those typically burned at the sites under normal operating conditions. Fired wastes varied widely at the different plants, with liquid organic and aqueous wastes most common. One incinerator, for example, had four waste feed streams: aqueous waste, liquid organic waste, and two types of drummed waste. Another incinerator burned five types of waste, which included one liquid and four solids. More specific descriptions of the individual burning procedures can be obtained in Trenholm et al.[2]

The hazardous organics present in the fired wastes at the eight sites include the following compounds.

- Methylene chloride
- Chloroform

- Methylene bromide
- 1,1,1-Trichloroethane
- Carbon Tetrachloride
- Trichloroethylene
- Benzene
- Tetrachloroethylene
- Toluene
- Chlorobenzene
- Hexachlorocyclopentadiene
- Chlordane
- Aniline
- *cis*-1,4-Dichloro-2-butene
- *trans*-1,4-Dichloro-2-butene
- Benzyl chloride
- Hexachlorethane
- Naphthalene
- Methyl ethyl ketone
- Phenol
- Butyl benzyl phthalate
- Bis (2-ethyl hexyl) phthalate
- Hexachlorobutadiene
- Phenyl isocyanate
- Phosgene

In addition to the type of organics present in the waste, key variables of the fired wastes included heating value and the percentages of chloride, ash, and moisture. Table 2 shows the variation of these waste characteristics, arranged by the waste types handled by each of the eight facilities.

III. SAMPLING PROCEDURES

Samples of the stack effluents were obtained by three principal methods during the trial burns: the EPA Modified Method 5 (MM5), gas bags, and the volatile organic sampling train (VOST).[2] Continuous monitoring of the stack emissions was also used at all the sites.

An MM5 train samples the stack emissions in the center of the stack, measuring for both particulate and organic emissions. For volatile principal organic hazardous constituent (POHC) sampling in stack effluent streams, two methods were used: (1) gas bags and (2) fast and slow VOSTs, a method developed by EPA for the trial burns. The gas bag samples were collected in 15-l bags into which stack effluent was drawn over approximately 1-h periods. In the fast VOST method, gas was drawn through two traps at a rate of 1 l/min for 20 min. The slow VOST, used in conjunction with fast VOST at some sites, drew stack gas through the traps at a rate of 0.25 l/min.

IV. RESULTS

Determination of DRE is the primary objective of a trial burn. In conjunction with DRE measurements for POHCs, however, products of incomplete combustion (PICs), hazardous metals, hydrogen chloride emissions, and carbon emissions were measured.

A. Destruction and Removal Efficiencies (DREs)

As discussed in Chapter 2, DREs for hazardous waste compounds relate the percentage of each substance that is destroyed and removed, and, consequently, the percentage of the

Table 2
SUMMARY OF WASTE FEED CHARACTERISTICS AT THE EIGHT TEST BURN FACILITIES[2]

Plant	Waste stream	Run no.	Heating value (kJ/kg)	Ash (%)	Chlorides (%)	Moisture (%)
Ross	Drum waste to kiln	1	18,680	36.9	0.019	27
		2	21,810	40.0	1.6	19
		3	22,100	41.0	0.12	14
	Drum waste to chamber	1	23,750	0.55	7.6	2.3
		2	19,010	0.51	7.5	8.4
		3	23,830	0.25	6.7	7.3
	Aqueous waste	1	1,570	0.25	0.04	98
		2	1,560	0.25	0.04	94
		3	1,510	0.26	0.03	95
	Liquid organic	1	21,890	2.42	6.2	24
		2	20,150	2.12	5.6	21
		3	22,230	2.28	5.4	23
Plant B	Liquid organic	1	19,760	4.9	5.4	35
		2	20,400	4.8	6.6	38
		3	19,710	4.7	5.4	48
	Aqueous	1	4,090	1.2	0.50	94
		2	4,480	1.1	0.57	95
		3	4,520	1.0	0.41	97
Upjohn	Liquid	2	23,770	0.17	21	b
		3	23,500	0.19	22	b
		4	23,980	0.21	21	b
	Gaseous	2	8,200[a]	0	132[b]	0
		3	8,200[a]	0	168[b]	0
		4	8,200[a]	0	239[b]	0
Zapata	Liquid organic	1	37,540	0.018	2.7	0.68
		2	37,890	0.013	1.6	0.63
		3	22,800	0.0098	1.3	37
		4	15,230	0.14	0.74	54
American Cyanamid	Aniline waste	1	33,700	0.16	0.018	6.0
		2	33,750	0.19	0.015	5.2
		3	33,670	0.19	0.020	5.5
		5	33,620	0.50	0.019	7.3
	Mononitrobenzene waste	4	25,060	<0.05	0.013	0.57
Mitchell	Aqueous	1	1,810	0.70	0.39	89
		2	1,700	0.78	0.36	94
		3	1,530	0.74	0.26	90
		4	1,600	0.80	0.38	90
	Organic	1	34,120	1.5	1.0	2.2
		2	34,400	2.1	1.3	3.2
		3	37,090	2.4	0.72	5.7
		4	34,650	0.78	1.2	2.4
DuPont	Liquid organic	1	29,110	0.47	25	2.8
		2	28,900	0.22	25	1.8
		3	28,910	0.30	26	2.9
	Filter coke	1	8,740	17	8.3	55
		2	8,430	17	9.8	51
		3	7,990	18	9.2	56
	DCB coke	1	28,710	0.55	22	0.47
		2	28,090	0.21	25	2.7
		3	27,560	0.72	21	4.8
	ISOM cake	1	15,930	0.68	18	39
		2	14,560	1.5	17	38
		3	19,030	1.6	19	32

Table 2 (continued)
SUMMARY OF WASTE FEED CHARACTERISTICS AT THE EIGHT TEST BURN FACILITIES[2]

Plant	Waste stream	Run no.	Heating value (kJ/kg)	Ash (%)	Chlorides (%)	Moisture (%)
	Paint waste	1	32,380	1.4	0.050	0.59
		2	31,830	2.3	0.18	0.43
		3	30,720	8.2	0.14	0.51
TWI	Liquid organic	1	35,060	3.1	1.7	4.9
		2	34,700	3.5	1.7	4.7
		3	34,150	3.9	1.8	5.3
		4	32,170	4.0	4.9	5.6
	Aqueous	1	158	<0.05	0.0097	94
		2	1,220	<0.05	0.013	94
		3	1,920	<0.05	0.0096	96
		4	1,920	<0.05	0.013	95
	Solids	1	810	90	2.0	1.5
		2	790	90	2.6	1.4
		3	670	90	2.0	1.6
		4	20,040	29	5.2	11

Note: Moisture analyses were not conducted. The moisture content is expected to be near zero because TDI was present in the waste in high concentrations, and TDI is highly reactive with water.

[a] kJ/dscm.
[b] mg/dscm.

compound that escapes through the incinerator's effluent streams. The minimum desired DRE for most compounds is 99.99%, meaning that 0.01% escapes to the environment in stack gas. The amount of POHCs leaving the incinerator through the stack gas is the primary determinant of the health risk derived from burning those compounds.

A number of incinerator performance factors were examined in the trial burns (excluding the preliminary trial burn in Cincinnati) to determine their impact, if any, on DREs: (1) waste feed concentration, (2) heat of combustion, (3) operating conditions including temperature and residence time, and (4) mechanisms separate from combustion, such as control equipment activity and air leakage. The relationships of these four factors to DRE, as examined at the eight facilities, are summarized below. For more details, see Trenholm et al.[2]

1. With increasing waste concentrations, there was a strong trend of increased DREs for both volatile and semivolatile compounds. At low waste concentrations (i.e., less than 100 μg/g), volatile compounds did not reach a DRE of 99.99%.
2. The trial burn data exhibited no positive correlation between heats of combustion and DRE measurements. Heat of combustion, alone, is therefore not an indicator of a compound's incinerability. This result casts doubt on EPA's use of the POHC system to determine incinerator efficiency.
3. Four incinerator operating conditions were analyzed with respect to DRE: residence time, chamber temperature, heat input rate, and the O_2 concentration in the incinerator stack. The only operating condition indicated by the data as having any significant influence on DRE was chamber temperature.
4. The air pollution control system at facilities may add chloroform and other trihalomethanes in small quantities to the stack effluents, thereby decreasing DREs. Any POHCs added to the stack emissions from inleak air, which enters the incinerator beyond the combustion chamber activity, may also decrease DREs.

Table 3
RANGE OF DRE RESULTS AND FRACTION (μg/g) OF UNDESTRUCTED POHCs AT THE CINCINNATI MSD INCINERATOR[3]

POHC	DRE (%)	μg of Undestructed POHC/g of POHC in the feed
	Runs 1 through 6	
Chloroform	99.998—99.9997	20—3
Carbon tetrachloride	>99.9[a]—99.999	<1,000—10
Tetrachloroethylene	>99.97[a]—99.999	<300—10
Hexachloroethane	>99.99—>99.999	<100—<10
Hexachlorobenzene	>99.99—>99.999	<100—<10
Hexachlorocyclopentadiene	>99.96[b]—>99.9999	<400—<0.2
	Runs 7 through 9	
Trichloroethane	99.985—<99.99996	150—<0.4
Tetrachloroethane	>99.9997—>99.9998	<3—<2
Bromodichloromethane	99.97—99.996	300—400
Pentachloroethane	99.9994—>99.9999	6—<1
Hexachlorothane	>99.999—>99.9999	<10—<1
Dichlorobenzene	>99.99—>99.999	<100—<10

[a] DRE could only be expressed as <99.9% or <99.97% because an inadequate sample was available for analysis from the integrated gas bag.
[b] DRE could only be expressed as >99.96% because the concentration of compound in one waste feed sample was quite low (90 ppm).

B. Principal Organic Hazardous Constituents (POHCs)

Of the 12 POHCs selected for DRE analysis at the Cincinnati MSD incinerator, all achieved a 99.99% DRE except bromodichloromethane. Table 3 summarizes the DREs for each POHC in the series of runs for this preliminary trial burn project.

For the purposes of the trial burns conducted at the eight facilities, POHCs were defined as EPA Appendix VIII compounds found in waste streams at concentrations above 100 μg/g. Toluene, tetrachloroethylene, carbon tetrachloride, and trichloroethylene were the most commonly detected POHCs in the wastes at the eight sites.

DREs were calculated for all volatile and semivolatile compounds. (A volatile is any substance that will evaporate at a low temperature.) Figure 1 shows the distribution of DREs for 134 volatile and 106 semivolatile data points. (A data point is one test run for one compound.) The average DRE for all volatiles at the eight incinerators was 99.992%. The semivolatile average was not determined, as more than half the stack measurements were below detection level limits. Tables 4 and 5 illustrate DREs calculated at two plants for some volatile and semivolatile compounds.

Low DREs for 80% of the volatiles and 96% of the semivolatiles occurred simultaneously with either or both of the following factors: (1) low concentration of the compound in the waste (<100 μg/g) and (2) the formation of PIC compounds. As stated previously, the data indicate that a low DRE correlates with a low concentration in the waste.

C. Products of Incomplete Combustion (PICs)

In the Cincinnati MSD incinerator trial runs, a PIC was defined as an Appendix VIII compound in the stack effluent that was not detected in the original waste. It is possible that a compound identified as a PIC in the stack gas may have been in the waste feed at a

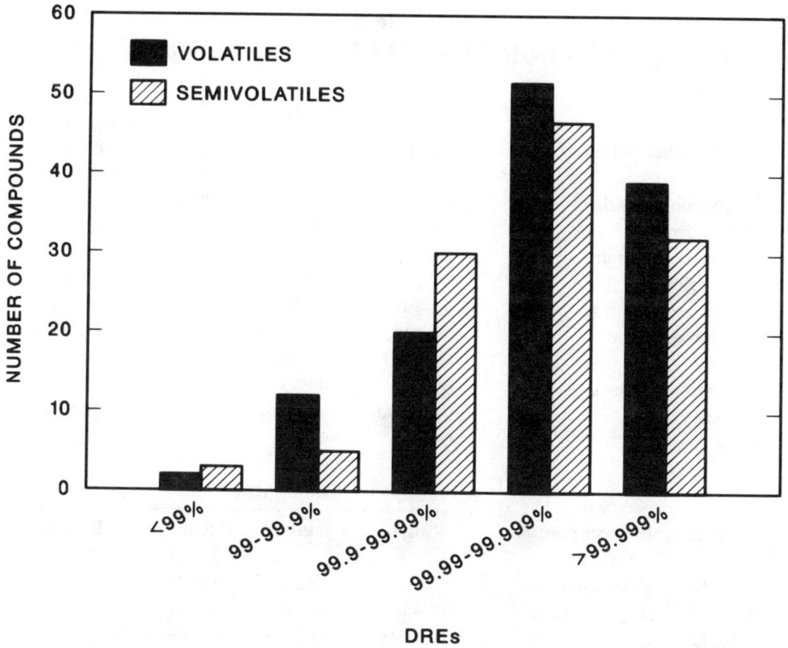

FIGURE 1. Destruction and removal efficiencies for volatile and semivolatile compounds at eight trial burns.

Table 4
DREs FOR THE ROSS INCINERATOR[2]

Volatile compounds	Fast VOST Run 1	Run 2	Run 3
Carbon tetrachloride	>99.9964	>99.9961	>99.9959
Trichloroethylene	>99.99936	99.9965	99.9969
Tetrachloroethylene	>99.9986	>99.9977	99.99912
Toluene	>99.99904	>99.9987	>99.9978
Methylene chloride	>99.968	>99.989	>99.978
Methyl ethyl ketone	99.99967	99.99930	99.99932
1,1,1-Trichloroethane	99.99952	>99.9990	>99.999
1,1,2-Trichloroethane	>99.999994	>99.999994	>99.999994

Semivolatile compounds	MM5 Run 1	Run 2	Run 3
N,N-Dimethyl acetamide	>99.9998	>99.9999	>99.9999
Phenol	>99.997	>99.993	>99.992
2,4-Dimethyl phenol	99.9992	99.9990	>99.9994
Naphthalene	>99.994	>99.994	>99.991
Butyl benzyl phthalate	>99.9996	>99.998	>99.999
Phthalic anhydride	[a]	>99.99	>99.99
Aniline	>99.998	>99.998	>99.998
Methyl pyridine	>99.998	>99.998	>99.998
Cresol(s)	>99.9993	>99.999	>99.991

[a] Waste feed concentration was <100 μg/g in this run.

Table 5
DREs FOR THE TWI INCINERATOR[2]

	Fast VOST			
Volatile compounds	**Run 1**	**Run 2**	**Run 3**	**Run 4**
Methylene chloride	>99.918	99.71	99.88	99.63
Chloroform	>99.944	98.2	97.8	99.78
Methylene bromide	>99.9987	99.9956	99.964	99.982
1,1,1-Trichloroethane	99.966	99.81	99.86	99.82
Carbon tetrachloride	>99.9984	>99.9983	>99.9987	>99.9990
Trichloroethylene	>99.9962	99.9945	99.9917	>99.9989
Benzene	99.9983	99.989	99.984	99.988
Tetrachloroethylene	99.965	99.78	99.88	99.982
Toluene	99.99946	99.9908	99.9963	99.9922
Chlorobenzene	99.965	99.70	99.956	99.966
	MM5			
Semivolatile compounds	**Run 1**	**Run 2**	**Run 3**	**Run 4**
Hexachlorocyclopentadiene	>99.99	>99.99	>99.99	>99.9996
Bis(2-ethyl hexyl)phthalate	99.951	99.960	99.940	99.88
Chlordane	>99.9998	>99.9999	>99.9999	[a]
Naphthalene	[a]	[a]	[a]	99.996
Hexachlorobutadiene	[a]	[a]	[a]	>99.98

[a] Waste feed concentration was <100 μg/g in this run.

level below detection limits (<100 ppm). The two PICs selected at the Cincinnati incinerator for calculation purposes were tribromomethane (bromoform) and dibromochloromethane. When the PIC emission rate was divided by the POHC input rate, PIC emissions did not exceed the proposed EPA regulation of 0.01% of POHC input.

The trial burns following the Cincinnati burn define a PIC as any Appendix VIII compound detected in the stack gas which was not detected in the waste in concentrations greater than 100 μg/g. Table 6 lists PIC concentrations in the eight sites' stack effluents for 29 compounds. Compared levels of PIC and POHC concentrations are found in Figure 2, and Table 7 lists PIC output as the percentage of POHC input for the eight facilities. In only one of the eight tests were PIC output rates greater than 0.01% of POHC input rates.

The presence of PICs in the stack gas was explained by three possible mechanisms: (1) low concentrations of the compound (<100 μg/g) were present in the waste and destroyed at a low DRE, (2) a source other than the waste introduced the compounds to the system, and (3) the PICs were products of combustion reactions. A comparison of these three contributing mechanisms to total PIC emissions is presented in Figure 3, which includes the percentage of the six most common PIC compounds. Chloroform, benzene, naphthalene, *O*-nitrophenol, bromoform, and phenol are listed as the most common PIC compounds found at the eight sites, based on frequency of occurrence and concentration in stack emissions.

Two families of PICs which are considered extremely toxic, dioxins and furans, were tested for in stack emissions at six incinerators. Three facilities emitted detectable furans, tetrachlorodibenzofurans (TCDF) and pentachlorodibenzofurans (PCDF). One incinerator was found to have effluent dioxins, tetrachlorodibenzodioxins (TCDD) and pentachlorodibenzodioxins (PCDD), in the stack gas. Stack sample concentration ranges (in nanograms per liter) for the detected dioxins and furans were: TCDF, 0.0007 to 0.047 ng/l; PCDF, 0.0005 to 0.023 ng/l; TCDD, 0.021 ng/l; and PCDD, 0.001 ng/l.

Plants having the most benzenes and/or phenols in the waste emitted the detectable dioxins

Table 6
PICs FOUND IN STACK EFFLUENTS[2]

PIC	Number of sites	Concentrations (ng/l)
Benzene	6	12—670
Chloroform	5	1—1,330
Bromodichloromethane	4	3—32
Dibromochloromethane	4	1—12
Bromoform	3	0.2—24
Naphthalene	3	5—100
Chlorobenzene	3	1—10
Tetrachloroethylene	3	0.1—2.5
1,1,1-Trichloroethane	3	0.1—1.5
Hexachlorobenzene	2	0.5—7
Methylene chloride	2	2—27
o-Nitrophenol	2	25—50
Phenol	2	4—22
Toluene	2	2—75
Bromochloromethane	1	14
Carbon disulfide	1	32
Methylene bromide	1	18
2,4,6-Trichlorophenol	1	110
Bromomethane	1	1
Chloromethane	1	3
Pyrene	1	1
Fluoroanthene	1	1
Dichlorobenzene	1	2—4
Trichlorobenzene	1	7
Methyl ethyl ketone	1	3
Diethyl phthalate	1	7
o-Chlorophenol	1	2—22
Pentachlorophenol	1	6
2,4 Dimethyl phenol	1	1—21

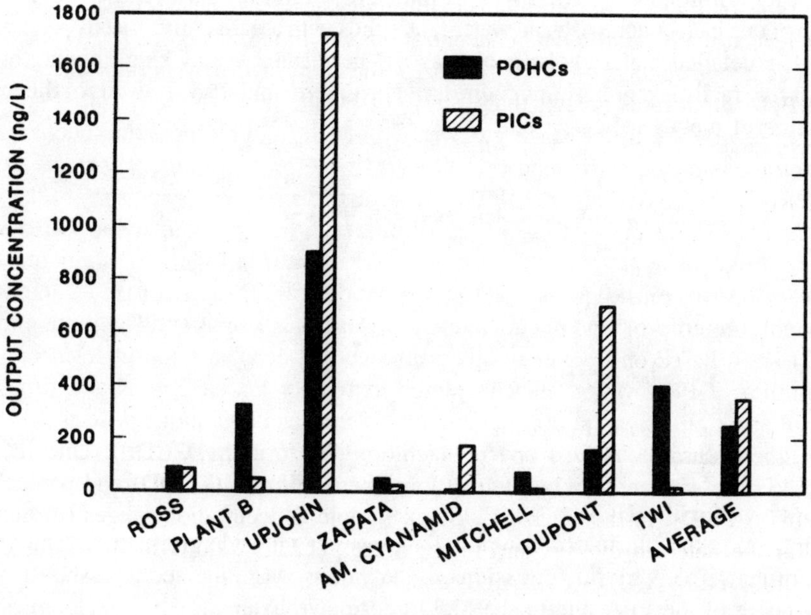

FIGURE 2. Comparison of PIC and POHC output at eight trial burn facilities.

Table 7
CALCULATED VALUES FOR PIC EMISSIONS AS A PERCENT OF POHC INPUT[2]

Plant	PIC output (g/min)	POHC input (g/min)	%[a]
Ross	0.090	9800	0.0009
Plant B[b] (semivolatiles only)	0.0046	100	0.0044
Upjohn	0.092	780	0.012
Zapata	0.00062	27	0.0023
American Cyanamid	0.011	1200	0.0009
Mitchell	0.00090	310	0.0002
DuPont	0.26	8300	0.0031
TWI	0.0076	1200	0.0006

[a] $\% = \frac{\text{PIC output (g/min)}}{\text{POHC input (g/min)}} \times 100$

[b] High blank values are for volatile PICs.

and furans, suggesting the two compounds as possible precursors of these PIC groups. (For more detailed treatment of PICs including dioxins and furans, see Chapter 7 of this book.)

D. Metals

Twelve metals are identified in EPA's Appendix VIII as hazardous constituents — antimony, arsenic, barium, beryllium, cadmium, chromium, lead, mercury, nickel, selenium, silver, and thallium. Metal emissions were analyzed at five of the eight trial burn facilities. The results are summarized in Table 8.

The data contained in Table 9 show the metals percentages of the waste feeds that were emitted in stack vapor. The percentages suggest that plants with high efficiency particulate control devices can significantly reduce hazardous metals output.

Due to the lack of data on metals concentration in incinerator ash and control device effluents, only a limited evaluation was formulated in the trial burn report. Table 10 gives an example of one incinerator's ash and water tank solids with regard to five metals percentages.

Although a detailed materials balance was not performed at any single incinerator, data collected at five facilities for barium, cadmium, chromium, and lead allowed for the following generalization of metals balance.

	Input Metals (%)	
Incinerator with particulate control	**Av**	**Range**
Metals emitted as particulate	3	0—27
Metals removed by scrubber	40	Not available
Metals in bottom ash	57	Not available
Uncontrolled incinerator		
Metals emitted as particulate	43	0—60
Metals in bottom ash	57	Not available

E. Hydrogen Chloride (HCl)

The combustion of chlorinated compounds generates chloride emissions, principally hydrogen chloride (HCl). Chloride emissions in hazardous waste incinerator exhaust gases are a major concern of the EPA due to the highly corrosive nature of HCl. HCl emissions are

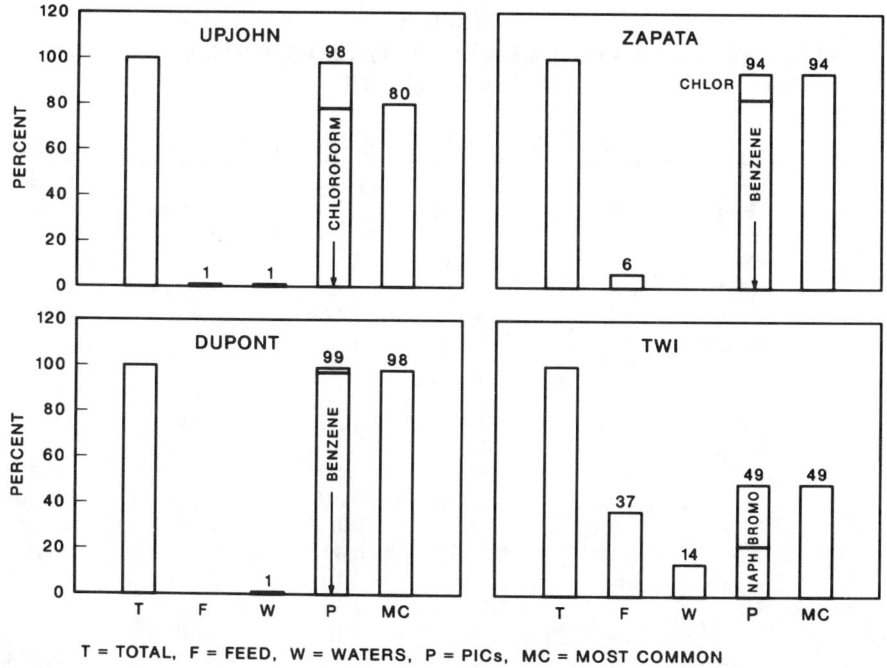

A

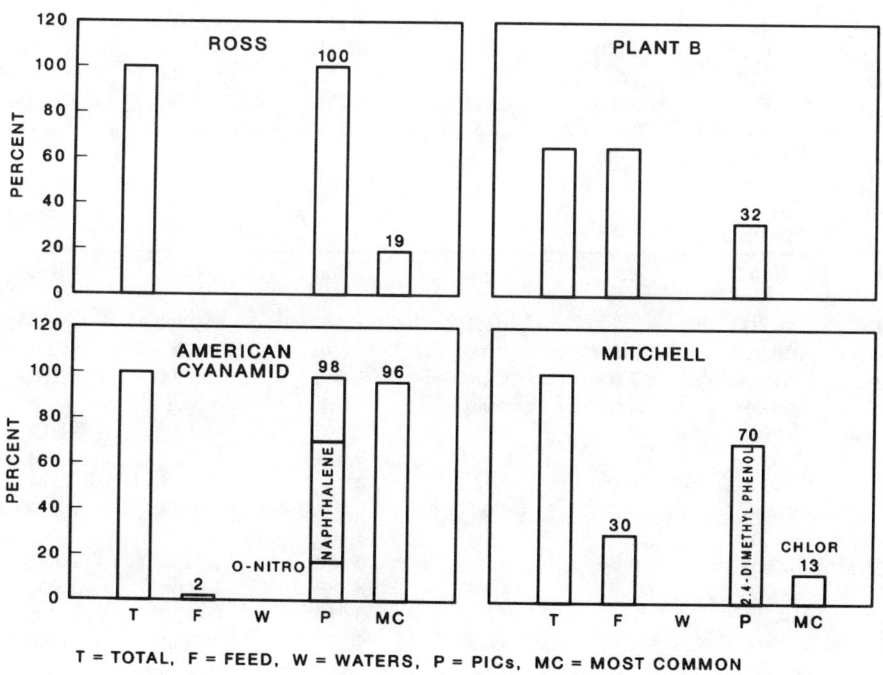

B

FIGURE 3. Comparison of mechanisms contributing to PIC emissions at eight trial burn facilities.

Table 8
SUMMARY OF PARTICULATE METALS EMISSIONS AT FIVE INCINERATORS[2]

	Ross			Plant B		American Cyanamid
Metal	Run 1	Run 2	Run 3	Run 2 inlet	Run 2 outlet	run 3
Antimony	820[a]	9,130	15,200	300	<2,500	<12,000
	0.063[b]	0.84	1.2	0.17	<0.019	0.48
Arsenic	<300	<240	<300	<72	<650	—
	<0.023	<0.022	<0.024	<0.035	<0.050	—
Barium	1,250	590	690	250	930	<1,000
	0.096	0.054	0.052	0.12	0.072	<0.04
Beryllium	<0.40	<0.32	<0.41	5	<43	—
	<0.00003	<0.00003	<0.00003	<0.002	<0.003	—
Cadmium	1,130	630	900	140	150	<1,200
	0.087	0.058	0.069	0.069	0.012	<0.005
Chromium	2,620	1,610	2,770	950	1,500	47,500
	0.20	0.15	0.21	0.47	0.12	0.19
Lead	91,600	68,900	96,100	3,100	5,300	<4,000
	7.1	6.4	7.3	1.5	0.41	<0.016
Mercury	56	<44	<56	<75	<680	<16,000
	0.0043	<0.041	<0.043	<0.037	<0.052	<0.064
Nickel	590	200	420	650	740	49,000
	0.045	0.018	0.032	0.32	0.57	0.20
Selenium	<600	<500	<600	9,200	3,700	—
	<0.046	<0.046	<0.047	4.5	0.29	—
Silver	14.8	11.7	7.6	<22	<200	620
	0.001	0.001	0.006	<0.011	<0.015	0.0025
Thallium	150	120	150	<120	<1,100	<10,000
	0.012	0.011	0.011	<0.059	<0.085	<0.004
Total hazardous metals	9.8—9.9[c]	8.1—8.2	11.6—11.7	1.5	1.2—1.7	9.7—14.1

DuPont			TWI			
Run 1	Run 2	Run 3	Run 1	Run 2	Run 3	Run 4
<1,200	<4,100	<1,700	<6,200	<2,800	<6,700	<8,800
<0.012	<0.012	<0.012	<0.06	<0.06	<0.06	<0.06
<5,000	<17,000	<7,000	<2,000	900	<2,200	<2,800
<0.049	<0.049	<0.049	<0.02	0.019	<0.02	<0.02
648	3,090	742	770	41	480	480
0.0064	0.089	0.0051	0.0072	0.0087	0.0051	0.0032
<120	<420	<180	<12	6.2	<12	<16
<0.0012	<0.0012	<0.0012	<0.0001	<0.00013	<0.0001	<0.0001
1,260	1,160	953	4,000	3,800	4,300	3,700
0.012	0.0033	0.0066	0.036	0.080	0.036	0.025
668	—	—	1,500	830	1,000	1,000
0.0066	—	—	0.014	0.017	0.087	0.0068
10,700	31,100	35,100	95,000	96,000	100,000	100,000
0.11	0.089	0.24	0.88	2.0	0.87	0.68
<1,000	<3,500	<1,500	<2,200	<960	<2,300	<3,100
<0.01	<0.01	<0.01	<0.02	<0.02	<0.02	<0.02
5,170	4,920	4,260	<460	230	<500	<660
0.051	0.014	0.030	<0.004	<0.004	<0.004	<0.004
13,800	61,600	28,300	<49,000	<22,000	<53,000	<70,000
0.14	0.18	0.20	<0.5	<0.5	<0.5	<0.5
648	1,880	222	<540	<240	<580	<770

Table 8 (continued)
SUMMARY OF PARTICULATE METALS EMISSIONS AT FIVE INCINERATORS[2]

DuPont			TWI			
Run 1	Run 2	Run 3	Run 1	Run 2	Run 3	Run 4
0.0064	0.0054	0.0015	<0.005	<0.005	<0.005	<0.005
<900	<3,100	<1,300	<4,000	<1,800	<4,300	<5,700
<0.0089	<0.0089	<0.0089	<0.04	<0.04	<0.04	<0.04
3.3—4.1	10.4—11.2	7.0—8.3	10.1—16.6	10.2—13.9	10.6—17.5	10.4—19.7

Note: < indicates that the concentration of metal was below the detection limit.

[a] The unit of measure for all first row numerical entries is μg/g of particulate.
[b] The unit of measure for all second row numerical entries is g/min.
[c] The unit of measure is % of total particulate.

limited to the greater of 1.8 kg/h (4 lb/h) of 1% of the wastes total organic chloride content under current Federal hazardous waste incinerator regulations.

Because HCl is not burned during combustion, other mechanisms must be employed for incinerators to meet Federal statutes regarding HCl emissions. The major mechanisms to control HCl output consist of (1) limiting the input waste chlorine content or (2) using APCDs.

The Cincinnati MSD incinerator was equipped with pollution control. Five incinerators of the eight analyzed by Trenholm et al.[2] also had some type add-on APCD (absorption scrubber) for HCl emissions. These scrubbing systems function by absorbing the gaseous HCl into the liquid. The other three trial burn incinerators did not have add-on control equipment, making it necessary to limit chlorine level in the input waste.

Table 11 summarizes the chloride removal performance of the Cincinnati incinerator during the preliminary trial burn. It is important to note that while 99% removal was not achieved in any of the 12 runs, the sampling method used in this trial burn measured total chloride removal rather than HCl removal efficiency. When the glass wool plug used in the sampling method was removed (due to its capture of large quantities of HCl from the control device scrubbing solution, the glass wool plug led to higher measurements of HCl emissions than were accurate), achievement of 99% removal was then estimated in half of the 12 test runs.

The trial burns at the eight facilities provide more data on HCl removal. The three facilities without HCl control devices fire low chloride wastes under normal operating conditions and would, therefore, easily comply with the Federal HCl emission standard of 1.8 kg/h. During the trial burns, two of these uncontrolled incinerators met the HCl regulation; the facility that did not comply with the standard failed to do so as a result of additional chloride being spiked into the waste feed. Table 12 summarizes the removal performance of the three incinerators lacking HCl emissions control equipment.

Of the five HCl-controlled facilities, three consistently achieved greater than 99% HCl removal; one incinerator reached 99% HCl control in one third of the tests; and one plant achieved less than 99% HCl control in three of the four test runs. The plants with repeatedly less than 99% HCl emissions control, however, were able to meet the Federal HCl emission standard by emitting less than 1.8 kg/h. Table 13 summarizes the performance of facilities equipped with HCl control systems.

F. Other Variables

To expand the incinerator emissions data base, stack effluents were continuously monitored for carbon monoxide and total hydrocarbons (THC), see Table 14. The trial burn results

Table 9
METAL EMISSIONS AS A PERCENTAGE OF METALS INPUT[2]

Site	Air pollution Control device	Run no.	Metals emissions/metals input (%) Sb	As	Ba	Be	Cd	Cr	Pb	Hg	Ni	Se	Ag	Tl
Ross	Yes	1	[a]	[a]	0.16	[b]	0.96	0.88	8.4	[a]	0.42	[a,b]	[a]	[a]
		2	5.3	[a]	0.13	[b]	1.1	0.73	9.7	[a,b]	0.05	[a,b]	[a]	[a]
		3	9.1	[a]	0.13	[b]	12.3	1.2	11.4	[a,b]	0.26	[a,b]	[a]	[a]
Plant B-Inlet	No	2	14—19[c]	[a,b]	6.3	[a,b]	[a]	56	63—71[c]	[a]	200	56—82[c]	[a]	[b]
Outlet	Yes	2	[a]	[a,b]	3.8	[a,b]	[a]	14	17—20[c]	[a]	36	3.6—5.6[c]	[a]	[b]
American Cyanamid	No	3	[a,b]	[b]	[a,b]	[a,b]	[a,b]	34	[b]	[b]	47	[b]	[a]	[b]
DuPont	Yes	1	[b]	[a]	1.6	[a,b]	3.1	0.43	4.2	[a]	0.28	[a,b]	[a]	[a,b]
		2	[b]	[a]	13.3	[a,b]	0.94	[b]	1.2	[a]	0.091	[a,b]	[a]	[a,b]
		3	[b]	[b]	0.75	[a,b]	2.4	[b]	1.8	[a]	0.25	[a,b]	[a]	[a,b]
TWI[c]	Yes	1	[a,b]	[a,b]	0.24	[b]	26	1.8	25	[a,b]	[b]	[a,b]	[a,b]	[a,b]
		2	[a,b]	[a]	0.26	[b]	50	1.9	51	[a,b]	[b]	[a,b]	[a,b]	[a,b]
		3	[a,b]	[a,b]	0.15	[b]	25	1.0	26	[a,b]	[b]	[a,b]	[a,b]	[a,b]
		4	[a,b]	[a,b]	0.24	[b]	37	1.8	43	[a,b]	[b]	[a,b]	[a,b]	[a,b]

Note: Figures are based only on liquid organic waste and solid waste feeds.

[a] Amounts in waste inputs contained < values.
[b] Amounts in particulates samples contained < values.
[c] Portions of waste inputs contained < values.

Table 10
METALS CONCENTRATION IN TWI CHAMBER ASH AND WATER TANK SOLIDS[2]

Metal	Run 1 Ash/leachate		Run 2 Ash/leachate		Run 3 Ash/leachate		Run 4 Ash/leachate		EP toxicity limit (mg/g)	Water tank solids
	mg/g	mg/l	mg/g	mg/l	mg/g	mg/l	mg/g	mg/l		
Barium	0.19	0.36	0.28	0.34	0.21	0.46	0.31	0.80	100	4.2
Cadmium	0.0021	0.016	0.0022	0.023	<0.002	0.0088	<0.0021	0.0084	1.0	0.21
Chromium	0.07	0.014	0.091	0.013	0.066	<0.007	0.18	0.015	5.0	0.73
Lead	0.081	0.12	0.11	0.10	0.094	0.058	0.079	<0.04	5.0	8.0
Nickel	0.012		0.02		0.011		0.024			<0.035

Table 11
SUMMARY OF CHLORIDE REMOVAL EFFICIENCY DATA AT THE CINCINNATI MSD INCINERATOR[3]

Date	Run no.	Cl in waste feed (%)	Total	Chloride removal efficiency (%) excluding plug
7/21/81	1	2.91[a]	95.3	97.7
7/21/81	1	2.91[a]	95.7	99.3
7/23/81	2	8.22	98.9	99.8
7/23/81	2	3.17	96.5	98.9
7/24/81	3	7.08[a]	98.4	99.7
7/24/81	3	7.08[a]	98.5	99.8
7/25/81	4	4.05	96.6	99.2
7/25/81	4	4.74	83.1	97.8
7/26/81	5	5.88[a]	93.3	97.7
7/26/81	5	5.88[a]	92.2	98.5
7/27/81	6	6.97[a]	96.6	99.7
7/27/81	6	6.97[a]	0	67.9

[a] The 6-h integrated waste feed sample was the only sample analyzed for Cl. Other analyses shown are for the 2-h waste feed samples.

Table 12
SUMMARY OF HCl EMISSIONS DATA FROM TEST BURN FACILITIES WITH NO AIR POLLUTION CONTROL SYSTEMS[2]

Site	Run no.	Chloride feed rate (kg/h)	Emission rate (kg/h)	Control efficiency (%)
Zapata	1	1.07	1.01	99.060
	2	0.74[a]	0.63	99.150
	3	0.63[a]	1.25	98.020
	4	0.34[a]	1.50	95.590
Mitchell	1	3.76	1.87	99.500
	2	4.27	2.23	99.480
	3	3.07		
	4	4.30	1.73	99.600
American Cyanamid	1	0.026	0.004	99.850
	2	0.015	0.002	99.870
	3	0.027	0.003	99.890
	4	0.025	0.003	99.880
	5	0.025	0.003	99.880

[a] Values are suspect due to nonrepresentative waste samples.

indicate that CO and THC are not good predictors of incinerator efficiency. Trenholm et al.[2] concluded that although CO concentration in the stack gases may indicate changes in incinerator performance, it does not help to predict POHCs or PICs in the stack, nor does it help to predict DRE.

Table 13
SUMMARY OF HCl EMISSIONS FROM TEST BURN FACILITIES WITH AIR POLLUTION CONTROL SYSTEMS[2]

Plant	Cl^- waste run no.	Cl^- emission feed rate (kg/h)	Control rate (kg/h)	Efficiency (%)
Ross	1	214	0.0678	99.968
	2	197	0.134	99.932
	3	167	0.132	99.921
Plant B	1	54.7	0.29	99.5
	2	69.8	0.83	98.8
	3	50.8	2.03	96.0
Upjohn	1	195	0.42	99.79
	2	274	0.53	99.81
	3	218	0.78	99.64
DuPont	1	287	0.235	99.99
	2	290	0.295	99.90
	3	305	0.407	99.87
TWI	1	7.8	0.135	98.3
	2	10.2	0.161	98.4
	3	7.8	0.251	96.8
	4	15.6	0.0978	99.4

V. CONCLUSIONS

The principal conclusions that can be reached from the EPA trial burn data are as follows:

1. DREs of 99.99% are achievable for POHCs. However, the trial burn data indicate that there is little relationship between a compound's heat of combustion and its incinerability. Current evidence seems to indicate that all organics are destroyed to essentially the same extent under a given set of conditions, regardless of their heat of combustion.[1] If this is the case, then the POHC system could provide an effective means of determining an incinerator's overall effectiveness in destroying organic compounds.
2. Metal emissions may be a larger problem at hazardous waste incinerators than emissions of POHCs. The trial burn data indicate that for incinerators with particulate control devices, about 3% of the metals in the waste feed are emitted in stack vapor. Incinerators without particulate control devices emit an average of 43% metals in the waste feed.
3. The characterization of PICs during the trial burns was inadequate. Less than 10% of the organics emitted from the incinerator stack have been identified. The EPA Science Advisory Board has recommended that the EPA characterize incinerator emissions more completely, specifying the identity of all chemicals released to the environment during the incineration process.[1]

Table 14
SUMMARY OF CONTINUOUS MONITORING DATA AT THE EIGHT FACILITIES[2]

Plant	Run no.	CO (ppm_v) Range	CO (ppm_v) Average	THC (ppm_v) Range	THC (ppm_v) Average
Plant B	1	10.6—22.5	14.8	<1	<1
	2	<1	<1	<1	<1
	3	1.6—10.7	6.9	<1.0—1.9	<1
	4	5.9—8.4	7.2	<1.0—1.1	<1
	5	3513—5400	4300	266—523	341
Ross	1	2.1—7.4	4.8	<1	<1
	2	4.7—15.5	9.1	0.8—2.3	0.9
	3	2.9—8.7	4.7	<1—2.3	1.0
Upjohn	1	10.3—11.0	10.5	7.1—11.9	8.9
	2	10.7—11.4	11.2	4.5—9.0	6.0
	3	9.3—10.0	9.9	3.1—6.2	3.9
Zapata	1	739—1570	1,275	32—215	71
	2	<1—394	22.2	<1—40.9	1.9
	3	3.6—8.8	7.5	<1	<1
	4	2.8—18.3	8.8	<1—2.9	<1
American Cyanamid	1	1.2—30.7	6.7	<1—1.6	<1
	2	8.3—37.3	19.3	<1—2.6	<1
	3	—	—	<1—1.2	<1
	4	7.0—25.5	13.8	<1	<1
	5	8.6—25.7	14.3	<1—1.1	1.0
Mitchell	1	0—4.6	1.4	<1	<1
	2	0.2—3.0	1.8	<1	<1
	3	0—12.7	<1	0.2—1.8	0.6
DuPont	1	34—1150	666	45.0—140	75.9
	2	34—1510	422	36.2—85.8	47.6
	3	60—1510	624	39.4—86.3	58.1
TWI	1	1.2—66.3	4.3	2.0—2.9	2.5
	2	0—13.8	0.9	1.7—2.1	1.9
	3	0—2.3	1.2	1.3—2.2	1.7
	4	0—2.0	0.6	0.3—2.1	0.8

REFERENCES

1. Assessment of Incineration As a Treatment Method for Liquid Organic Hazardous Wastes, Background Report I: Description of Incineration Technology, Office of Policy, Planning and Evaluation, U.S. Environmental Protection Agency, Washington, D.C., March, 1985.
2. **Trenholm, A., Gorman P., and Jungclaus, G.,** Performance Evaluation of Full-Scale Hazardous Waste Incinerators, *Vols. I-V,* EPA-600/2-84-181a-e, Industrial Environmental Research Laboratory, U.S. Environmental Protection Agency, Cincinnati, OH, November, 1984.
3. **Gorman, P. G. and Ananth, K. P.,** Trial Burn Protocol Verification at a Hazardous Waste Incinerator, Industrial Environmental Research Laboratory, U.S. Environmental Protection Agency, Cincinnati, OH, 1982.

Chapter 4

RISK ASSESSMENT METHODOLOGY

I. INTRODUCTION

Risk assessment is a procedure used to synthesize all available data and the best scientific judgment to estimate the risk associated with exposure to a chemical.[1] Risk estimates are expressed as a probability of harm. In numerical terms, the probability of harm can have a range of values between zero and one: zero risk indicates certainty that harm will not occur, while a risk value of one declares certainty that harm will occur.[2] Many risk calculations, such as the risks involved in driving a car, are based upon years of collected data and, as such, are quite accurate. Chemical exposure risks, however, including those derived from hazardous waste incineration, must be estimated from a limited data base, making the margin of error in the estimation much larger.

Historically, risk analyses to determine the health impacts from chemical exposures have been concerned with short-term effects such as dizziness and nausea.[1] Recently, however, the focus has shifted toward long-term adverse health effects, with an emphasis on lifetime cancer risks from exposure to toxic substances. The major obstacle to this type of risk assessment is the limited data base available on the carcinogenic effect of chemicals and the heavy reliance on animal studies whose results must be extrapolated across species to humans. The current status of risk analysis requires an abundance of scientific value judgments.

The first half of this chapter comprises a description of the present methodology employed to determine health risks from chemical exposures, including the inherent assumptions involved in the process. The second part describes the specific application of risk assessment procedures to hazardous waste incineration.

II. COMPONENTS OF RISK ASSESSMENT

Risk assessment, the characterization of potential adverse health effects of human exposures to environmental hazards, involves four consecutive steps:[3]

1. Hazard identification — determination of whether a chemical poses a potential health hazard to humans
2. Exposure assessment — determination of the extent of human exposure to the toxic chemical(s)
3. Hazard assessment — determination of the relationship between the magnitude of exposure and the probability of occurrence of health effects
4. Risk characterization — combination of the first three steps to estimate the health risk associated with human exposure to the substance under investigation

This sequence of steps is viable whether the health risk being evaluated is cancer or another endpoint.[1] We will now discuss each component briefly.

A. Hazard Identification

The purpose of the hazard identification step is to collect and evaluate information concerning the inherent toxic properties (potential to cause biological injury, disease, or death under certain exposure conditions) of chemicals. Four approaches can be employed to determine chemical toxicity: epidemiological studies, long-term animal bioassays, short-term tests, and comparisons of molecular structure. Data from these sources are reviewed to determine if a chemical is likely to be toxic to humans. The determination of toxicity is strengthened by consistency of data across test methods.[1]

1. Epidemiological Studies

Epidemiology is a branch of research concerned with the incidence of disease among groups of people, including the study of health impacts from chemical agents. When available, epidemiological studies can supply direct carcinogenicity data for human exposure to chemicals.[1] Such studies can correlate cancer incidence with exposure to industrial processes and factors pertaining to individual lifestyle as well. Epidemiological research, however, is limited to the study of disease derived from natural or accidental causes; researchers have little control over experimental parameters.

Analytic epidemiological studies can be divided into two primary types: cohort and case-control studies. Cohort studies compare two populations which are differently exposed to a substance; case-control studies contrast people having a particular disease with people who do not have that disease. These major epidemiology categories are often supported by case histories and descriptive studies. Although a given investigation might not provide evidence of increased cancer risk for an exposed population, it may demonstrate an upper limit risk likely to be manifest in a similarily composed and exposed group.[1]

Due to the lack of available epidemiological data, the majority of information used in assessing potential carcinogenic effects in humans comes from long-term animal bioassays. Table 1 lists the chemicals and processes for which epidemiological and animal bioassay data lend evidence of carcinogenicity. Note that only 13 chemicals and five industrial processes have been identified as having sufficient epidemiological evidence to be classified as definite human carcinogens. After evaluating all existing data in 1979, the International Agency for Research on Cancer (IARC) concluded that, at most, 54 chemicals and industrial processes could be considered as human carcinogens.[4] Since that time, other groups and individuals have expanded the list somewhat. As of 1985, however, the EPA had provided quantitative estimates of human cancer risk for only 54 chemicals.[5]

2. Long-Term Animal Bioassays

Long-term animal studies currently provide the largest and the most available data base for predicting potential human carcinogenicity of chemicals.[1] Increased tumor incidence in experimental groups is the primary health effect considered in animal bioassays. For the most part, carcinogenicity data relies on exposure testing for a major portion, or the duration, of an animal's lifetime (i.e., 2 to 3 years for rats and mice).[2] Doses at levels much higher than comparable human exposure doses are often administered to experimental animals to insure the production of enough cases of cancer and to assure statistical validity through the collection of enough data. The major difficulty with long-term animal bioassay data from a risk assessment point of view is the necessity to extrapolate the results (1) across species from rats or mice to humans and (2) from the high doses to which animals are exposed in the laboratory to the low doses to which humans are exposed in the environment.

3. Short-Term Tests

A third approach which can provide additional data for qualitative evaluation of a toxic substance involves short-term testing. Short-term tests include investigation of gene mutation, chromosome damage, DNA effects, and both *in vitro* and *in vivo* exposures. Animals in *in vivo* experiments, for example, may be exposed to potentially hazardous chemicals through such methods as skin painting. At the present time, these tests generally lend support or confirmation to long-term animal bioassay and epidemiological data in the hazard identification process.[1]

Table 1
EVIDENCE OF CARCINOGENICITY IN HUMANS AND ANIMALS FOR CHEMICALS AND INDUSTRIAL PROCESSES FROM IARC MONOGRAPHS[6]

Chemical or process	Degree of evidence: In humans	Degree of evidence: In experimental animals
1. Acrylonitrile	Limited	Sufficient
2. Aflatoxins	Limited	Sufficient
3. 4-Aminobiphenyl	Sufficient	Sufficient
4. Amitrole (aminotriazole)	Inadequate	Sufficient
5. Arsenic and certain arsenic compounds	Sufficient	Inadequate
6. Asbestos	Sufficient	Sufficient
7. Auramine	Limited	Limited
8. Manufacture of auramine	Sufficient	Not applicable[a]
9. Benzene	Sufficient	Inadequate
10. Benzidine	Sufficient	Sufficient
11. Beryllium and certain beryllium compounds	Limited	Sufficient
12. N,N-Bis(2-chloroethyl)-2-naphthylamine (chlornaphazine)	Sufficient	Limited
13. BCME and technical grade CMME	Sufficient	Sufficient
14. Cadmium and certain cadmium compounds	Limited	Sufficient
15. Carbon tetrachloride	Inadequate	Sufficient
16. Chlorambucil	Limited	Sufficient
17. Chloramphenicol	Inadequate	No data
18. Chlordane and heptachlor	Inadequate	Limited
19. Chloroprene	Inadequate	Inadequate
20. Chromium and certain chromium compounds	Sufficient	Sufficient
21. Cyclophosphamide	Limited	Sufficient
22. DDT	Inadequate	Limited
23. Dieldrin	Inadequate	Limited
24. Diethylstilbestrol	Sufficient	Sufficient
25. Dimethylcarbamoyl chloride	Inadequate	Sufficient
26. Dimethyl sulfate	Inadequate	Sufficient
27. Epichlorohydrin	Inadequate	Limited
28. Ethylene oxide	Limited	Inadequate
29. Hematite	Inadequate	Negative
30. Underground hematite mining	Sufficient	Not applicable[a]
31. HCH (technical HCH and lindane)	Inadequate	Limited
32. Iron dextran	Inadequate	Sufficient
33. Isoniazid	Inadequate	Limited
34. Isopropyl oils	Inadequate	Inadequate
35. Manufacture of isopropyl alcohol (strong acid process)	Sufficient	Not applicable[a]
36. Lead and certain lead compounds	Inadequate	Sufficient (for some soluble salts)
37. Melphalan	Sufficient	Sufficient
38. Mustard gas	Sufficient	Limited
39. 2-Naphthylamine	Sufficient	Sufficient
40. Nickel and certain nickel compounds	Limited	Sufficient
41. Nickel refining	Sufficient	Not applicable[a]
42. Oxymethalone	Limited	No data
43. Phenatecin	Limited	Limited
44. Phenobarbitone	Limited	Limited
45. *N*-Phenyl-2-naphthylamine	Inadequate	Inadequate
46. Phenytoin	Limited	Limited
47. Polychlorinated biphenyls	Inadequate	Sufficient
48. Reserpine	Inadequate	Inadequate
49. Soots, tars, mineral oils	Sufficient	Sufficient
50. Styrene	Inadequate	Limited
51. Trichloroethylene	Inadequate	Limited
52. Tris(aziridinyl)-*p*-benzo-quinone (triaziquone)	Inadequate	Limited

Table 1 (continued)
EVIDENCE OF CARCINOGENICITY IN HUMANS AND ANIMALS FOR CHEMICALS AND INDUSTRIAL PROCESSES FROM IARC MONOGRAPHS[6]

Chemical or process	Degree of evidence	
	In humans	In experimental animals
53. Tris(1-aziridinyl)phosphine sulfide (thiotepa)	Limited	Sufficient
54. Vinyl chloride	Sufficient	Sufficient

Note: Sufficient = "sufficient evidence to support a causal relationship between the exposure and cancer."

[a] It is difficult to expose experimental animals to the same conditions to which workers are exposed; therefore, no animal data are available.

4. Comparisons of Molecular Structure

A fourth method of hazard identification involves comparison of the structure of the chemical of concern with that of known carcinogens.[3] The presumption is that a chemical which is structurally similar to a known carcinogen has a high probability of being a carcinogen itself. Although there is some evidence supporting such associations for a few chemical classes, at present these studies are best used to identify chemicals for a more thorough investigation and set priorities for carcinogenicity examination.

5. Assumptions in Hazard Identification

The four classes of data discussed above are reviewed to determine if a chemical is likely to pose a health hazard to humans. The greater the consistency of data across test methods, the greater the confidence in such a prediction. Because of the lack of human epidemiological data and the difficulties in interpreting short-term test results, the burden of proof in the hazard identification step usually lies on data obtained from long-term animal bioassays.[1]

The methods employed in identifying potential chemical carcinogens are based on scientific data, but reasonable assumptions and scientific judgments must be made to compensate for the lack of complete information. Some primary assumptions and judgments involved in the hazard identification process are as follows:[1]

- The agent administered in an animal bioassay is in a form similar to that present in the environment.
- Animal carcinogens are also human carcinogens.
- There is a similarity in animal and human metabolism.
- The dose received by the target site is directly proportional to the administered dose.

B. Exposure Assessment

The second component in risk assessment is exposure analysis. Obviously, there is no risk without exposure. The exposure assessment component in risk analysis attempts to define the route, magnitude, frequency, and duration of exposure of a given population to a toxic substance.[1] The principal pathways of exposure generally considered in exposure assessment are atmospheric transport, surface and groundwater transport, ingestion of toxic materials that have passed through the aquatic and terrestrial food chain, and dermal absorption (Figure 1). Because of the complexity of the processes involved in the environmental distribution of chemicals, exposure assessment is often the most resource-demanding portion of an evaluation of chemical risks.[1] "Many authorities agree that the weakest link in our understanding of environmental health studies is our knowledge of human exposure."[7]

Once an exposure assessment determines the quantity of a chemical with which human

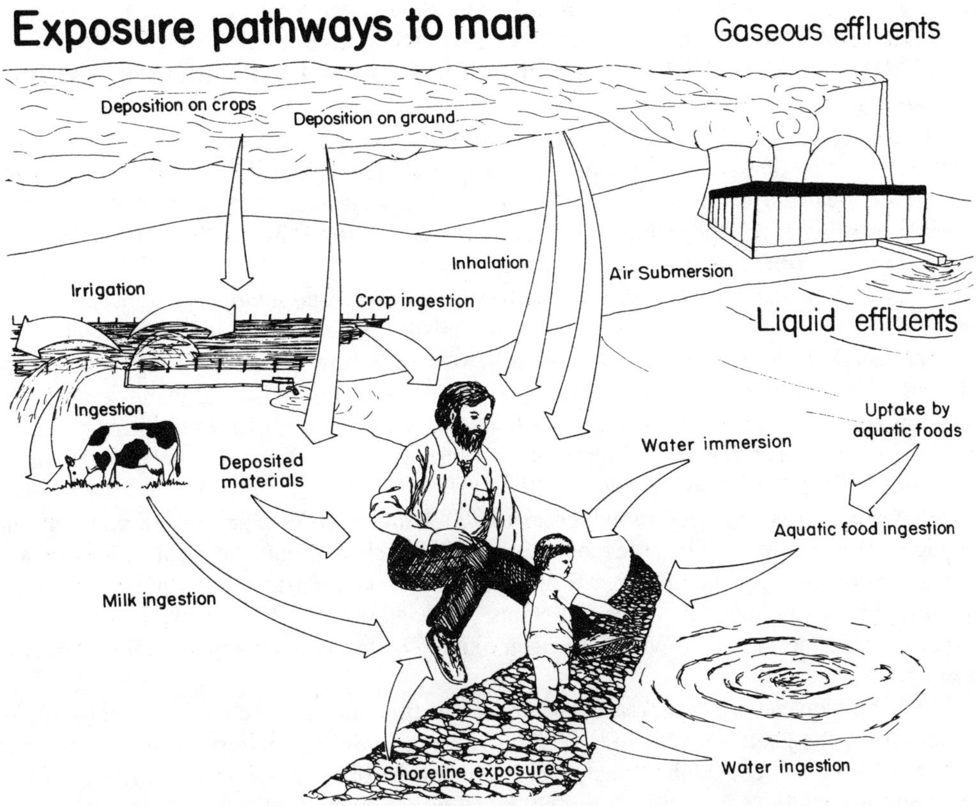

FIGURE 1. Exposure pathways to man from incinerator emissions.

populations may come in contact, the information can be combined with toxicity data (from the hazard identification process) to estimate potential carcinogenic risks. There are three basic methods for determining the concentration of a pollutant to which target populations are exposed. The first is direct measurement of the pollutant concentration in the environment. This is a common practice in occupational settings and in some urban environments. The second is direct measurement of biological indicators in the human body to provide an indirect estimate of exposure. This is a new and rapidly developing field. In some situations, however, either concentrations are too low to be detected or direct measurement is costly or impossible. Under these circumstances, the third method, mathematical models describing environmental transport, is used to predict environmental concentrations of pollutants.

In most instances, a combination of site monitoring data and environmental modeling is required to produce the best estimates of chemical concentrations at points of exposure.

1. Monitoring Studies

Monitoring provides information on individual exposures and correlates sources of chemical release to human exposure levels.[1] It can be used to study chemical sources, ambient concentrations, chemical exposures in the home and on the job, exposures through food and water consumption, and the doses absorbed and retained in the body. Some disadvantages are involved, however, with the collection of data from personal monitors. Detection levels of personal monitors, for example, may limit some exposure investigations. Presently, the majority of personal monitors give data for average exposures, but do not provide information regarding exposure times and peak levels. In addition, monitoring for chemical exposures in the work place is often site specific and fails to estimate total exposures for the entire work force population.[1]

2. Modeling

Mathematical models can estimate human exposure levels where monitoring data are either unavailable or incomplete. These models are best utilized when the estimates desired are average or idealized exposure levels. Because models are simplifications of the depicted exposure phenomenon, the exposure predictions derived from this method are often crude estimations, regardless of the models' sophistication.

The basic processes that control the environmental fate of pollutants are transport, transformation, and intermedia transfer. Transport is the movement of pollutants within the environmental media of air, water, and soil, due to natural forces. During atmospheric transport, for example, the movement of pollutants is controlled by ambient winds. The direction and speed of winds determine where the center of mass of an airborne pollutant will be found. In surface and groundwater, transport mechanisms are the movements of water and sediment carrying chemicals.

Transformation is any process which causes a change to take place in the physical and chemical structure of an airborne pollutant. Terms such as photochemical oxidation and biodegradation refer to specific transformation processes. Transformation may involve interaction with particulates or other pollutants. Perhaps the best known example is photochemical smog formation, in which nitrogen oxides, hydrocarbons, and sunlight interact to form ozone.

Transformation can either increase or decrease the biotoxicity of a chemical. For example, atmospheric photochemical reactions are known to increase the toxicity of the smoke from residential wood stoves. Of the three processes that control the environmental fate of chemicals and, ultimately, human exposure, chemical transformation is the least understood.

Intermedia transfer is the movement of pollutants between the environmental media of air, water, soil, and biota. For example, an airborne pollutant can be transferred from the atmosphere to the earth's surface through gravitational settling or washout by rain. A pollutant in a river may transfer to the atmosphere through volatilization, or it may absorb to suspended sediment and deposit out of the river. A pollutant in soil may volatilize to the atmosphere or leach into groundwater. Cross-media transfer can result in a wide environmental distribution of pollutants and, consequently, greater potential for human exposure from multiple sources.

3. Atmospheric Transport

The primary pathway of human exposure to pollutants release during hazardous waste incineration is inhalation. It is therefore necessary to estimate the downwind concentrations of pollutants (see Figure 2). The Gaussian plume model is a widely used method of estimating downwind concentrations of airborne material released to the atmosphere. This class of models is based on the assumption that the plume will spread both laterally and vertically in accordance with a normal, bell-shaped statistical distribution. The models have been successfully used for many years to predict ground-level air concentrations where reasonably flat terrain exists and where an average, fairly uniform release rate can be assumed. It has generally been found that near-field Gaussian plume models are able to predict annual-average atmospheric concentrations of nonreactive pollutants to within a factor of two to four over flat terrain.[8]

Most air quality models use diffusion coefficients and eddy diffusivity developed from studies over flat, homgeneous terrain. In areas of mountain ranges and deep valleys, wind and temperature fields produce complicated atmospheric dispersion patterns and distributions that are difficult to simulate. These include: (1) channeling of flows along valleys, (2)

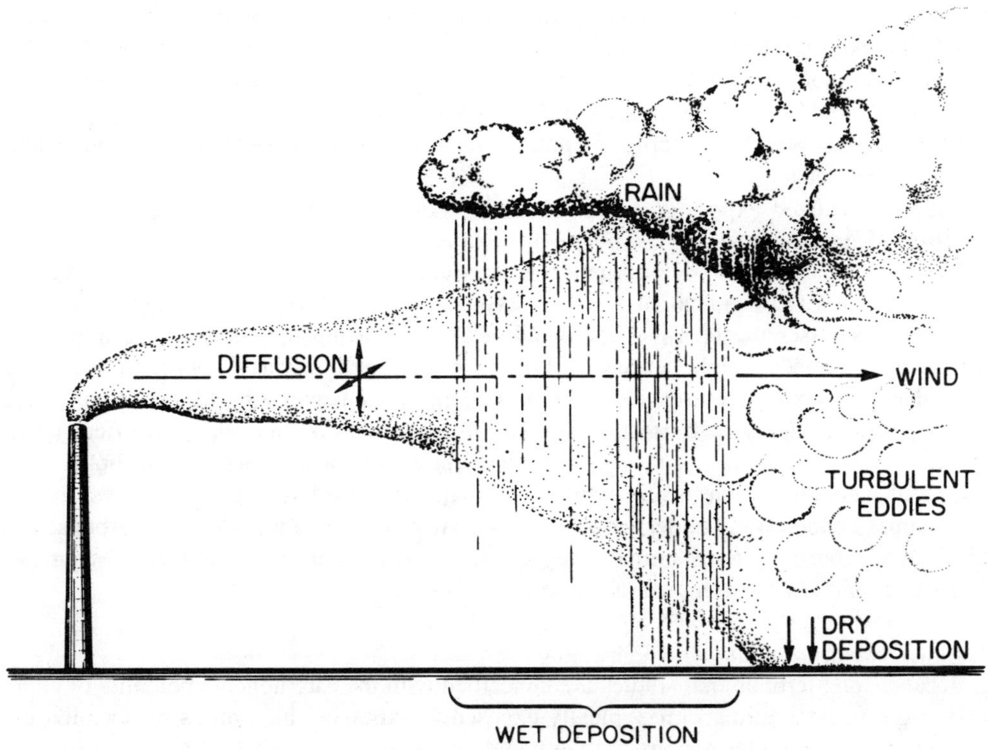

FIGURE 2. Illustration of downwind concentrations of pollutants.

trapping of pollutants within basins, (3) enhanced vertical dispersion through mechanically induced turbulence, (4) large vertical gradients in wind speed and direction, and (5) flow changes resulting from heating and cooling of sloping terrain. Under conditions of complex terrain, Gaussian plume models are less applicable. Short-term concentration predictions under conditions of complex meteorology or terrain may be uncertain by one or two orders of magnitude.[9]

4. Model Uncertainty

Uncertainties exist in exposure models because they are mathematical approximations of the real world. These uncertainties arise from two sources: (1) imprecision in the description of the physical and chemical processes and (2) uncertainties in parameters used as input to the models. The best method of determining model uncertainty is through comparison of model predictions with independent field or laboratory observations made over the range of conditions for which application of the model is intended.[8] Despite a recent emphasis on model testing and validation, unified and accepted procedures and measures for model validation do not exist at the present time.[10,11]

5. Human Activity Patterns

In determining human exposure to a chemical, the behavior of the human population under study should be taken into consideration, whether the exposure study employs monitoring, modeling, or a combination of data-gathering approaches.[1] Variability in and uncertainty of an exposure value for a population can result when people change jobs, relocate residence, convert from using well water to municipal water, and take medication, etc.[1] Daily activity patterns also contribute to variability in exposure levels. In a study of two New Jersey cities, individuals were continuously monitored for exposure to 20 atmospheric organic chemicals.[12] Due to changes in activity patterns throughout the day and night, monitored exposure levels fluctuated greatly with indoor exposures of all compounds greatly exceeding outdoor exposures. Thus, stationary monitoring of only outdoor pollutant concentrations would have underpredicted total exposure. Despite an awareness of the importance of human activity patterns, at present they are rarely accounted for when making exposure estimates.

C. Hazard Assessment

Hazard assessment, the determination of expected human response to a given exposure level, is the third component of a risk assessment and is referred to in some literature as dose-response assessment.[1] Dose is typically expressed as milligrams of chemical received per kilogram of body weight per day (milligrams per kilogram per day). For toxic substances other than carcinogens, a threshold dose must be exceeded before a health effect will occur. Threshold doses apply to those toxic chemicals for which a "no observed effect level" (NOEL) exists. For cancer-causing substances, however, an absence of thresholds is assumed. A carcinogenic hazard assessment is usually presented as a specific risk estimate or as an upper bound on the underlying risk. The majority of carcinogenic dose-response data are derived from long-term animal study information. Mathematical models are then used to convert animal data to human applicability.

1. Low Dose Extrapolation

Because long-term animal studies are concerned with the carcinogenic potential of chemicals, experimental animals are typically exposed to extremely high doses of potential carcinogens (at or near their maximum tolerated dose level) in order for the cancers to develop. These doses are generally several orders of magnitude higher than those encountered by humans in the environment. In order to employ the results of long-term animal studies in a risk assessment for expected human exposure levels, some form of extrapolation from the high-dose range to the low-dose range (using mathematical models) is required. Low-dose extrapolation may also be necessary for risk assessments based on epidemiological studies, since doses resulting from accidental or occupational toxic exposures for humans may be relatively high as well.[1]

In addition to high dose levels, animal bioassays produce high risk measurements due to the small number of animals used in the laboratory.[2] If five animals out of the 50 exposed develop cancer, then the risk at that dose is 10%. This is an extremely high risk measurement, but it is about the smallest amount of risk that can be detected in average-sized animal studies. The application of mathematical models is necessary to predict human risks from such high dose/high risk animal studies. Low-dose extrapolation through modeling attempts to describe the little known relationship between exposure and human response.[1]

2. Mathematical Models

Three general categories of models can be utilized to characterize the dose-response relationship: (1) tolerance distribution models, (2) mechanistic models, or (3) time-to-tumor models.[1] Tolerance distribution models assume that each individual possesses a threshold or tolerance level with respect to developing cancer following exposure to a carcinogen. When an individual's dose level is above the threshold, he will develop cancer following exposure. When it is below the threshold, he will not develop cancer. The difficulty with this approach is the lack of knowledge concerning individual threshold levels. To circumvent this difficulty, it is assumed that a probability distribution can describe the variability among individual threshold levels. The probability that a member of a population will exhibit a carcinogenic response when exposed at dose level d is then equal to the probability that the same individual's threshold level is less than d.

The mechanistic class of models includes the one-hit, multistage, and multihit models, which are based on a presumed understanding of carcinogenic mechanisms. These models are designed to reflect the notion that single cells, when damaged by a chemical or one of its metabolites, result in tumor formation. The majority of biological data support this class of models, and it is the class frequently used in risk assessment. The linearized, multistage model employed by the EPA in its risk assessments if of this class.

The third model category, time-to-tumor models, includes the lognormal distribution, Weibull distribution, and the Armitage-Doll and Hartley-Seilkin models. These are the most sophisticated low-dose extrapolation models. This class attempts to describe the relationship between toxic dose, tumor latency, and cancer risk and may lead to a more complete characterization of the carcinogenic process. However, applicability of time-to-tumor models may be limited by insufficient data. Previous evidence also suggests that precision of low-dose risk estimation does not increase due to implementation of more sophisticated models.[1]

Criticism of testing chemicals at high doses and the ensuing low-dose extrapolation process revolves around the argument that metabolism of chemicals differs at high and low doses. Some scientists assert that high doses "may overwhelm normal detoxification mechanisms and provide results that would not occur at the lower doses to which humans are exposed."[3] Another point of controversy is the unknown discrepancy between the administered dose and the dose of carcinogen which is received by the target tissue or organ. At present, since tissue or organ dose is unknown, all dose-response estimates must be based on the administered dose.

Most dose-response models are quantitatively similar in the range of observable response rates (10 to 100%) (see Figures 3 and 4), yet they commonly provide estimates that vary by three to four orders of magnitude at lower, unobservable response rates.[13] It is these low, unobservable carcinogenic response rates which are of interest to regulatory agencies (i.e., 1 in 100,000). Presently, one model cannot be selected over the others as yielding better estimates, due to a lack of knowledge regarding specific validity. Animal bioassay data are also not sufficient to isolate any of the competing models as the best method.

3. Interspecies Extrapolation

Once a low-dose risk estimate is obtained through modeling, this number must still be extrapolated to the human population.[1] Differences between the animal species tested and the human population that may affect their respective responses to carcinogenic exposures include: life spans, body size, genetic variability or population homogenicity, presence of disease, and pharmacokinetic effects such as metabolism.

Interspecies extrapolation has traditionally used standardized dosage conversions such as (1) milligrams per kilogram of body weight per day (milligrams per kilogram per day), (2) milligrams per surface area per day (milligrams per meters^2 per day), and (3) parts per million in diet or water per exposed portion of life (parts per million per lifetime). It is still not known whether laboratory animals are relevant surrogates for the human situation; however, current data suggest that the best agreement between predicted and observed human cancer risk results from the use of the milligrams per kilogram of body weight per day or the milligrams per surface area per day approaches to interspecies extrapolation. The assumption is that when dose is expressed in one of these two measures, effects across species will be approximately the same. Although the choice of dose conversion scale affects the magnitude of the risk calculation,[1] none of the above conversion approaches has been validated biologically.[13]

4. Assumptions in Hazard Assessment

As with all phases of the risk assessment process, hazard assessment is based on many assumptions. In addition to the previously discussed assumptions involved with modeling

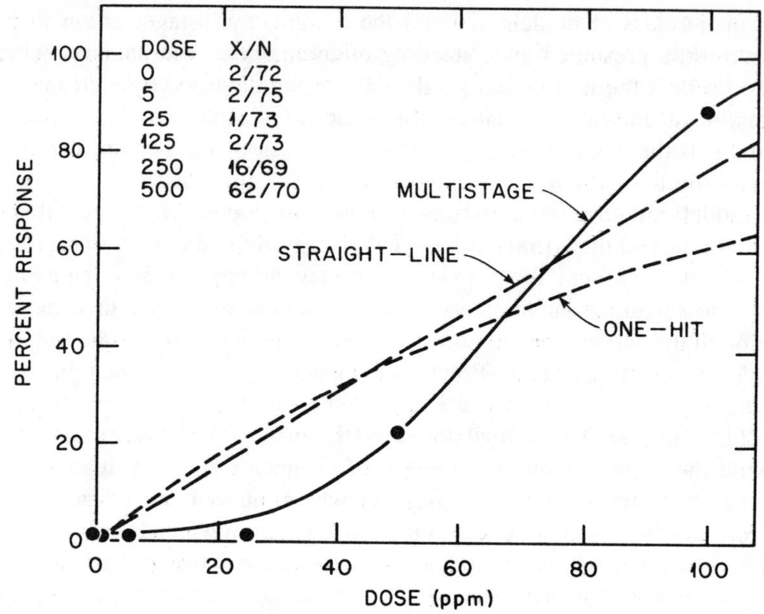

FIGURE 3. Incidence of thyroid carcinoma in rats exposed to ethylene thiourea. (Modified from Reference 13.)

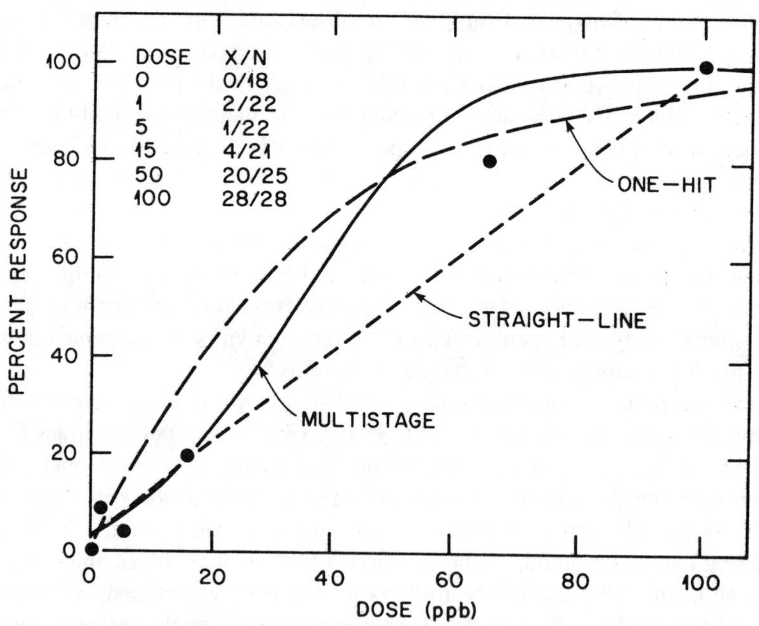

FIGURE 4. Incidence of liver tumors in rats exposed to aflatoxin B_1. (Modified from Reference 13.)

for low-dose and interspecies extrapolation, hazard assessment does not allow for such factors as (1) different dose-response levels between children, adults, and the elderly, (2) different physiological factors within populations, and (3) biochemical activity between the carcinogenic agent and the site where it is received.[1]

D. Risk Characterization

The final component in risk assessment methodology is risk characterization, which involves compilation and evaluation of all relevant data. Risk is then estimated as the probability that population members will likely evidence adverse health effects from exposure to the substance in question.[1] As this integration is based on all the assumptions required to complete the three preceding steps, it is important for a risk assessment to discuss the level of uncertainty inherent to the final estimates.

III. RISK ASSESSMENT FOR HAZARDOUS WASTE INCINERATION

In estimating the human health risk from hazardous waste incineration, two measures of risk are usually employed: maximum individual risk and population risk.

Maximum Individual Risk (MIR) — MIR is measured as the maximum risk to an exposed person. It quantified the probability of developing cancer following a lifetime (70 years) of continuous exposure to a given contaminant.

Population Risk (PR) — Population risk measures the impact of a pollutant on the entire population exposed. It quantifies the total number of cancers per year which can be expected in a population as the result of exposure to a given contaminant.

These estimates of human health risk from hazardous waste incineration are calculated using the following procedures:

1. The rate of release (micrograms per second) of specific chemicals in incinerator stack exhaust is either measured directly or estimated from incinerator operating parameters.
2. Ground-level atmospheric concentrations of pollutants are determined within a given radius of the incineration facility using stack emission rates and atmospheric transport models.
3. These atmospheric concentrations (micrograms per meter^3) are then multiplied by the number of persons inhabiting the area surrounding the incinerator to quantify the population exposure (person — micrograms per meter^3) and the maximum individual exposure (micrograms per meter^3).
4. The maximum individual exposure and the population exposure are multiplied by a conversion factor, which takes into account the average weight, breathing rate, and absorption of the contaminant by individuals in the population to yield the maximum individual dose (milligrams per kilogram per day) and the population dose (person — milligrams per kilogram per day).
5. The maximum individual dose and the population dose estimate are multiplied by the chemical's unit potency factor, a measure of cancer risk resulting from a dose of 1 mg/kg of body weight per day for 70 years. Thus, the two primary estimates of cancer risk, MIR and PR, for exposure to a chemical are obtained.

A more detailed description of this procedure follows.

A. Exposure Assessment

The first major step in the risk assessment procedure for hazardous waste incinerators is the determination of population exposures and doses (steps 1 through 4 above). We briefly describe the process.

1. Stack Emissions

The rate of release (mass emissions per unit of time) of specific chemicals in incinerator stack exhaust is controlled by three facility variables: (1) waste throughput, (2) chemical concentration in the waste stream, and (3) destruction and removal efficiencies (DRE).

Under normal operating conditions, the stack emission rate for organics is given by:[14]

$$Q = Q_1 f_1 f_2 \tag{1}$$

where Q = emission rate of uncombusted chemical, grams per second; Q_1 = waste throughput, grams per second; f_1 = fraction of waste stream represented by constituent of concern; f_2 = fraction of waste uncombusted based on 99.99% DRE = 1×10^{-4}.

2. Air Dispersion

Ground-level air concentrations of chemical pollutants are generally estimated via Gaussian plume atmospheric transport models (see the previous section on atmospheric transport). These dispersion models are applied for distances of 50 to 100 km around the site under investigation.

Data necessary for the operation of atmospheric transport models are incinerator stack parameters, pollutant behavior variables, and region-specific meteorological input. Incinerator stack parameters considered include stack height and diameter, exit gas temperature, exit gas velocity, and pollutant emission rate. Meteorological input spans such categories as wind speeds, air temperature, and mixing layer height.

3. Exposure Duration

The exposure duration depends on both the behavior patterns of the exposed population and various physical phenomena. The time a person lives in an area, commuting habits, and personal habits all affect exposure duration. Typically, such patterns are impossible to characterize objectively and are, therefore, simply neglected to be conservative. It is generally, therefore, assumed that exposure takes place continuously over an individual's entire lifetime.

The lifetime value used in an exposure assessment should represent the average life expectancy of the exposed population. Life expectancy in the U.S. has been steadily increasing over recent years. The overall average is now 74 years.[15] Most risk assessments, however, assume a 70-year lifetime.

4. Body Weight

The body weight used in an exposure assessment should reflect the average weight of the exposed population during the time when the exposure actually occurs. If the exposure occurs continuously throughout an individual's life or only during the adult ages, using the adult average weight provides sufficient accuracy. Under these circumstances, the body weight is typically assumed equal to 70 kg, which represents an average adult male.[16]

5. Absorption

Following inhalation or ingestion of a substance, a fraction is absorbed into the body. Absorption is a function of the exchange membrane properties, the route of administration, and the contaminant properties. Normally, the absorption value is based on data from animal studies. Applicable animal data should be collected in a manner which best simulates the human exposure scenario of concern. The EPA has produced two sets of references which contain absorption data for a variety of chemicals: (1) Health Assessment and Air Criteria Documents and (2) Water Quality Criteria Documents. These documents present respiratory tract absorption factors and gastrointestinal (GI) tract absorption factors. Under conditions where absorption data are unknown, it is common to assume 100% absorption. The absorption factor refers to the absorption into the body of contaminants entering the respiratory system.

6. Inhalation

The general equation for estimating inhalation exposure is as follows.

$$\text{Inhalation exposure} = \frac{\text{Inhalation rate} \times \text{Contaminant concentration in air} \times \text{Exposure duration} \times \text{Respiratory absorption factor}}{\text{Body weight} \times \text{Lifetime}} \quad (2)$$

Under conditions of a lifetime exposure, this equation becomes

$$\text{Inhalation exposure} = \frac{\text{Inhalation rate} \times \text{Contaminant concentration in air} \times \text{Respiratory absorption factor}}{\text{Body weight}} \quad (3)$$

The estimate of exposure dose obtained from this formula is in units of milligrams per kilogram per day. The inhalation rate varies according to the exertion level. These levels and associated times have been characterized by the International Commission on Radiological Protection (ICRP)[16] as (1) 7.5 l/min for 8 h/d of resting and (2) 20 l/min for 16 h/d engaged in light activity. Combining these values, a daily average inhalation rate of 22.8 m^3 is obtained. This value should be used when the exposure is continuous throughout the day. If the exposure is not continuous, the inhalation rate used in a risk assessment should be adjusted to reflect the portion of the day over which the exposure occurs.

7. Ingestion

For estimating ingestion exposure, the general equation is

$$\text{Ingestion exposure} = \frac{\text{Ingestion rate} \times \text{Contaminant concentration in food or water} \times \text{Exposure duration} \times \text{GI tract absorption factor}}{\text{Body weight} \times \text{Lifetime}} \quad (4)$$

The contaminant concentration refers to the concentration in food or whatever is being ingested. The absorption fraction refers to the absorption occurring in the GI tract.

Under conditions of a lifetime exposure, this equation becomes

$$\text{Ingestion exposure} = \frac{\text{Ingestion rate} \times \text{Contaminant concentration in food or water} \times \text{GI tract absorption factor}}{\text{Body weight}} \quad (5)$$

The estimate of exposure dose obtained from this formula is in units of milligrams per kilogram per day.

Some typical ingestion rates used by the EPA in risk assessment are listed in Table 2. They represent national averages derived from surveys. Ingestion rates for other food groups are available from the Food and Drug Administration (FDA).[17] These surveys also offer more detail on how ingestion rates vary according to location, age, and other parameters.

Table 2
AVERAGE U.S. INGESTION RATES[18,19]

Substance	Consumption rate
Water	2 l/d
Freshwater fish	6.5 g/d
Beef	124 g/d
Dairy products	550 g/d

Such information can be used to further tailor the exposure assessment to site-specific conditions.

GI tract absorption fractions are, in large part, derived from animal studies. As mentioned earlier, animal data should be selected which best matches the human exposure scenario. The vehicle in which the compound is administered to animals should correspond to the vehicle in which the contaminant contacts people. As an example, Poiger and Schlatter[20] find that the absorption of dioxin in ethanol is about twice as great as the absorption of dioxin in soil. Thus, it is important in a soil ingestion exposure calculation to use an absorption fraction derived from studies that use a soil vehicle.

B. Consequences of Exposure

Quantitative risk assessment emphasizes reliances on numerical results to evaluate health impacts. Consequences of exposure to hazardous chemicals emitted from incinerators can be estimated as carcinogenic and noncarcinogenic health effects. Following are descriptions of both types of risk and sample calculations for each.

1. Carcinogenic Chemicals

Based on animal and human data, the EPA has provided quantitative estimates of human cancer risk for 54 chemicals.[5] The cancer potency factor for a specific chemical estimates the largest possible linear slope at low doses that is consistent with the empirically derived dose-response curve. It is calculated using the improved multistage model developed by Crump[21] in the case of animal data and by the relative risk model in the case of human data.[22]

Excess lifetime risk factors developed by the EPA are interpreted as the probability, at the 95% confidence level, that an individual will develop cancer after a 70-year exposure to the chemical at a dosage of 1 mg/kg of body weight per day. "Excess" means that the risk is in addition to the risk of cancer presented by other sources. Table 3 presents the EPA excess cancer risk potency factors for 62 hazardous constituents for which values are available.[23] These factors are expressed in units of (milligrams per kilogram per day)$^{-1}$,

$$\text{Cancer risk} = \text{Potency (mg/kg/day)}^{-1} \times \text{Exposure (mg/kg/day)} \qquad (6)$$

A cancer potency factor of 1.0×10^{-3} (mg/kg/d)$^{-1}$, for example, indicates an excess lifetime cancer risk of one case per thousand individuals exposed to 1 mg/kg/d for 70 years. Cancer risk estimates may range between zero and one and represent the upper bound lifetime risk of dying of cancer as a result of a continuous 70-year exposure. The EPA's estimates of excess risk are purposefully conservative on the side of protecting human health. Thus, the true or expected risks associated with exposure to these chemicals are smaller than the calculated risks with 95% confidence. The cancer potency estimates presented in Table 3 demonstrate an extremely broad range of excess risk; seven orders of magnitude. Most of the risk factors, however, tend toward the lower end of the distribution. The notable exception

Table 3
CARCINOGEN POTENCY VALUES (INHALATION) FOR HAZARDOUS WASTE CONSTITUENTS[23]

Carcinogenic constituent	Cancer risk factor $(mg/kg/day)^{-1}$
Dioxin (2,3,7,8-tetrachlorodibenzo-*p*-dioxin)	425,000
Bis(chloromethyl)ether	9,300
Aflatoxins	2,930
1,2-Dimethylhydrazine	803
Dibenzo(a,i)pyrene	476
Benzidine	234
Diethylstilbestrol	107
N-Nitrosodiethylamine	43
4-Nitroquinoline-*l*-oxide	40
Coal tars	31
Dieldrin	30
N-Nitrosodimethylamine	25
Arsenic and compounds	14
Benzo pyrine	11.5
Aldrin	11.4
Hexachlorocyclohexane, Alpha isomer	11.1
Reserpine	10.4
Kepone	8.97
DDT	8.42
N-Nitrosodi-*n*-butylamine	5.43
Beryllium and compounds	4.86
Polychlorinated biphenyls	4.34
2,4-Diaminotoluene	4.03
Hydrazine	3.91
Heptachlor	3.76
Heptachlor epoxide (Alpha, Beta, Gamma)	3.37
N-Nitrosophyrrolidine	2.13
Hexachlorocyclohexane, Beta	1.84
3,3-Dichlorobenzene	1.69
Hexachlorobenzene	1.68
Chlordane	1.61
Lindane	1.33
Bis(2-chloroethyl)ether	1.14
Toxaphene	1.13
1,1-Dichloroethene	1.04
1,2-Diphenylhydrazine	0.77
Amitrole	0.74
Acrylonitrile	0.55
Methyl hydrazine	0.36
2,4-Dinitrotoluene	0.31
Pentachloronitrobenzene	0.26
1,1,2,2-Tetrachloroethane	0.20
Chloroform	0.18
Urethane-ethyl carbamate	0.15
Thiourea	0.14
Formaldehyde	0.083
Carbon tetrachloride	0.083
Hexachlorobutadiene	0.078
1,1,2-Trichloroethane	0.057
Benzene	0.052
1,1,2,2-Tetrachloroethene	0.040
1,2-Dichloroethane	0.037

Table 3 (continued)
CARCINOGEN POTENCY VALUES (INHALATION) FOR HAZARDOUS WASTE CONSTITUENTS[23]

Carcinogenic constituent	Cancer risk factor $(mg/kg/day)^{-1}$
Aniline	0.030
Safrole	0.025
2,4,6-Trichlorophenol	0.019
Vinyl chloride	0.017
Hexachloroethane	0.014
Ethylphthalyl ethylglycolate	0.014
Bis(2-ethylhexyl)phthalate	0.014
Trichloroethylene	0.013
1,4-Dioxane	0.010
N-Nitrosodiphenylamine	0.005

is dioxin which is the most potent carcinogen tested to date. Since this chemical is produced as a byproduct of incineration, it is the focus of discussion in several of the subsequent chapters.

2. *Sample Calculation for Carcinogenic Risk*

In Chapter 5, we estimate the carcinogenic risk associated with a hazardous waste stream in a sequence of steps. First, we determine the maximum atmospheric concentration to which an individual is exposed (micrograms per meter^3). This provides the maximum individual exposure. Then, we multiply the average atmospheric concentration (micrograms per meter^3) in each sector segment resulting from incineration by the number of individuals residing in that sector segment and add these numbers for all sector segments to quantify the population exposure (person — micrograms per meter^3). We then multiply maximum individual exposure and population exposure by a conversion factor, which takes into account the average weight and breathing rate of individuals in the population to yield maximum individual dose (milligrams per kilogram per day) and population dose (person — milligrams per kilogram per day). This, in turn, is multiplied by the excess cancer risk factor developed by the EPA to yield estimates of maximum individual risk and population risk.

For example, consider a situation in which a chemical is dispersed through the atmosphere resulting in a maximum individual exposure of 10 $\mu g/m^3$ and a population exposure of 1×10^6 person — micrograms per meter^3. If the chemical had an excess cancer risk factor of 1×10^{-2} $(mg/kg/d)^{-1}$, the calculation of risks is as follows:

$$\text{MIR} = (10\ \mu g/m^3)\ \frac{22.8\ m^3}{\text{person}-\text{day}}\ \frac{1\ mg}{1000\ \mu g}\ \frac{1\ \text{person}}{70\ kg}$$

$$(0.50)\ \frac{1 \times 10^{-2}\ \text{Cancers}}{mg/kg/day} = 1.6 \times 10^{-5} \qquad (7)$$

The 0.5 in the above calculation represents the amount of chemical assumed to be absorbed and retained upon inhalation. The 22.8 m^3/d represents the reference man's breathing rate, while the 70 kg represents the reference man's body weight.[16] Thus, the maximum individual risk is 1.6×10^{-5}. That is, the maximally exposed individual has a probability of 1.6 chances in 100,000 of developing cancer as a result of a continuous 70-year exposure to this chemical.

$$\text{Pr} = (1 \times 10^6 \text{ person} - \mu\text{g/m}^3)\frac{22.8 \text{ m}^3}{\text{person} - \text{day}}\frac{1 \text{ mg}}{1000\ \mu\text{g}}\frac{1 \text{ person}}{70 \text{ kg}}$$

$$(0.50)\frac{1 \times 10^{-2} \text{ Cancers}}{\text{mg/kg/day}} = 1.6 \qquad (8)$$

Thus, the population risk is 1.6 total excess cancer deaths in the exposed population over 70 years.

3. Noncarcinogenic Chemicals

One hundred fourteen chemicals have been identified by the EPA as noncarcinogenic constituents of hazardous waste streams. For these chemicals, the acceptable daily intake (ADI) is used as the measure of chronic toxicity. An ADI is defined by the EPA as the amount of a toxic agent in milligrams per kilogram of body weight per day (or in milligrams per day for a 70 kg person) which is not expected to result in any adverse effects after chronic exposure in the general population of humans, including sensitive subgroups.[23] ADIs are generally calculated by dividing the no-observed-adverse-effect level (NOAEL) derived from animal or human toxicity studies by an uncertainty factor to insure protection of human health. The recommended safety factors are 10, 100, or 1000 depending on the type of data used to determine the NOAEL:

- A safety factor of 10 is used when the NOAEL is based on chronic human data. This factor is meant to account for intraspecies variability.
- A safety factor of 100 is used when the NOAEL is based on chronic animal data. This factor is meant to account for both intra- and interspecies variability.
- A safety factor of 1000 is used when the NOAEL is based on subchronic animal data. This factor incorporates uncertainty in extrapolating from subchronic to chronic exposures and includes the two previous uncertainty factors.

Table 4 presents the EPA estimates of ADIs for the chemical constituents of currently incinerated waste streams. These chemicals possess ADIs ranging over five orders of magnitude. The chemicals with the smallest ADIs are the most toxic, while the chemicals with the largest ADIs are the least toxic.

4. Sample Calculation for Noncarcinogenic Risk

For noncarcinogens, a measure of health risk is obtained by comparing the estimated daily intake of the chemical with the ADI. The accepted daily intake is considered safe for humans; at this lifetime exposure level, it is very unlikely that a toxic response will manifest.[2] If the estimated daily intake is less than the ADI, there should be no adverse effects from continuous exposure to the chemical. As an example, consider an individual exposed to an average concentration of 1 μg/m^3 of a chemical with an ADI of 0.1 mg/d. The daily intake as a fraction of ADI is computed as follows:

$$1\ \mu\text{g/m}^3 \times 22.8\frac{\text{m}^3}{\text{day}} \times \frac{1 \text{ mg}}{1000\ \mu\text{g}} \times 0.5 \times \frac{1}{0.1\ \mu\text{g/day}} = 0.15 \qquad (9)$$

The 22.8 m^3/d represents the reference man's breathing rate and the 1/0.1 μg/d the inverse of the ADI. Thus, under the above conditions, an exposed individual would receive 15% of the ADI.

Table 4
ACCEPTABLE DAILY INTAKES (INHALATION) FOR HAZARDOUS WASTE CONSTITUENTS, SORTED BY ASCENDING ADI[23]

Noncarcinogenic constituent	ADI (mg/day)
Phorate	0.0024
Lead (Pb)	0.0045
4-Chlorophenol	0.0050
Acrolein	0.0090
Nitroglycerine	0.0095
Toluene diisocyanate	0.013
Sodium azide or smite	0.014
Silver and compounds	0.016
Phosgene	0.019
Methyl parathion	0.020
Mercury	0.020
Cyanogen chloride	0.029
Hexachlorocyclopentadiene	0.036
Thallium and compounds	0.037
Bromodichloromethane	0.039
Parathion	0.040
1,1-Dimethylhydrazine	0.048
Maleic anhydride	0.048
Endrin	0.070
Bis(2-chloroisopropyl)ethyl	0.070
4,6-Dinitro-*o*-cresol and salts	0.071
2,6-Dinitrotoluene	0.071
Hydrofluoric acid	0.071
Methyl ethyl ketone peroxide	0.071
Oxirane	0.095
Acrylamide	0.11
Methomyl	0.12
Chloroacetaldehyde	0.14
1,3-Dichloropropene	0.18
Pentachlorophenol	0.18
Benzyl chloride	0.24
Dinitrobenzene	0.28
1,3-Dinitrobenzene	0.28
1,2-Dinitrobenzene	0.28
1,4-Dinitrobenzene	0.28
Endosulfan	0.28
Crotonaldehyde	0.29
Phthalic anhydride	0.29
Antimony and compounds	0.29
Cyanides (soluble salts)	0.33
1,2,4,5-Tetrachlorobenzene	0.35
Diphenylamine	0.36
Tetranitromethane	0.38
Furfural	0.38
Tribromomethane	0.39
Benzo(j,k)fluorene	0.42
Formic acid	0.43
1,2,4-Trichlorobenzene	0.46
2-(Chloromethyl)oxirane	0.48
Hydrocyanic acid	0.52
Selenium and compounds	0.70
Pyridine	0.71

Table 4 (continued)
ACCEPTABLE DAILY INTAKES (INHALATION) FOR HAZARDOUS WASTE CONSTITUENTS, SORTED BY ASCENDING ADI[23]

Noncarcinogenic constituent	ADI (mg/day)
2,4-Dimethylphenol	0.80
DDE	0.84
Dimethylamine	0.86
1,2-Dichloropropane	0.98
Chlorobenzene	1.01
Methoxychlor	1.40
Carbon disulfide	1.43
Acrylic acid	1.43
Bromomethane	1.45
Nickel and compounds	1.46
Ethylbenzene	1.60
Methyl ethyl ketone	1.72
Trichloronaphthalene	1.80
Di-*N*-butyl-phthalate	1.80
Dimethyl phthalate	1.80
Resorcinol	2.14
DDD	3.01
Nitrobenzene	4.00
Cyclohexanone	4.76
Hydrogen sulfide	5.00
Methyl methacrylate	5.23
2,6-Diaminotoluene	5.70
Phenol	6.80
2,4-Dichlorophenol	7.00
2,4,5-Trichlorophenol	7.00
1-Butanol	7.14
Isobutyl alcohol	7.14
Cresols	7.90
Acetaldehyde	8.57
2,4-Dinitrophenol	9.45
Methyl isobutyl ketone	9.76
Isophorone	10.50
Methanol	12.38
Dichloromethane	12.90
N-Phenylthiourea	13.70
Pentachlorobenzene	14.00
Zinc and compounds	15.00
Dichloropropane	16.66
Naphthalene	18.00
Xylene	20.71
Acetonitrile	25.00
Tetrahydrofuran	28.09
Acetophenone	29.60
2,5-Diaminotoluene	30.30
Chloromethane	37.50
1,1,1-Trichloroethane	37.50
Cyclohexane	50.00
Ethyl ether	57.14
Ethyl acetate	66.66
Acetone	84.76
1,2-dichlorobenzene	107.00
Tetrafluoromethane	129.00

Table 4 (continued)
ACCEPTABLE DAILY INTAKES (INHALATION) FOR HAZARDOUS WASTE CONSTITUENTS, SORTED BY ASCENDING ADI[23]

Noncarcinogenic constituent	ADI (mg/day)
Toluene	134.00
Chlorotrifluoromethane	153
Dichlorobenzene	161
1,4-Dichlorobenzene	161
Dichlorofluoromethane	177
Trichlorofluoromethane	201
Butylphthalyl butylglycolate	700
Diethyl phthalate	875
1,1,2-Trichloro-1,2,2-trifluoroethane	1910
Monochloropentafluorethane	11280

IV. ACCEPTABLE RISK

"The knowledge of risk magnitudes is of little use . . . unless . . . related to the magnitudes of everyday risks, for otherwise no sense is given of the importance of the risks and how much notice should be taken of them."[24] We briefly review risk magnitudes presented by two categories of risk: everyday activities and risk levels of chemical carcinogens which have led to federal regulation.

A. Everyday Risks

One way to obtain a perspective of the risk levels associated with hazardous waste incineration is by comparing them to risk levels associated with other human activities. Table 5 presents risk of death over a lifetime (70 years) from several voluntary and involuntary exposures. These data indicate that about 8 out of every 1000 individuals will die in a home accident during their lifetime and that approximately 2 out of 100 (2%) of all individuals will die in an automobile accident. About 2 out of every 10 individuals (20%) will die of some form of cancer.

Most of the risks in Table 5 are familiar ones that individuals tend to worry little about. However, some scientists consider it inappropriate to compare risks from incineration with those associated with everyday activities. How one evaluates a given risk level depends on many factors, including whether the risk is voluntary or involuntary and whether the risks result from new activities or from long-established practices.

A more appropriate measure of acceptable risk might be levels of additional risk that individuals are willing to tolerate from environmental carcinogen exposure. The best way to determine such levels is to review, retrospectively, the use of cancer risk estimates in setting prevailing Federal standards.

B. Regulation of Chemical Carcinogens

Before reviewing risk levels that have led to regulation of chemical carcinogens, we define two terms: *de manifestis* and *de minimis*. A *de manifestis* risk is a risk level of obvious or evident concern. In legal terms, an "obvious risk" is defined as one that is "instantly recognized by a person of ordinary intelligence". With regard to environmental carcinogens, we use the term to denote a level of risk that has always resulted in regulatory action to reduce the risk.

The term *de minimis* is used to define an acceptable level of risk: a level considered safe

Table 5
RISKS OF DEATH FROM VOLUNTARY AND INVOLUNTARY EXPOSURES[24,25]

Source of exposure	Lifetime average risk[a]
Accidents	
All home accidents	7.7×10^{-3}
Lightning	3.5×10^{-5}
Fire arms	7.0×10^{-4}
Motor vehicles	1.7×10^{-2}
Electrocution	3.7×10^{-4}
Cancers	
All cancers	1.9×10^{-1}
Buccal cavity, pharynx, and respiratory	5.0×10^{-2}
Digestive organs and peritoneum	5.3×10^{-2}
Bone, connective tissue, skin and breast	2.2×10^{-2}
Genital organs	2.2×10^{-2}
Urinary tract	8.4×10^{-3}
Leukemia and other blood and lymph	1.8×10^{-2}
Radiation	
Airline pilot (50 hr/month at 12 km altitude)	2.8×10^{-3}
One transcontinental round trip by air per year	7.0×10^{-5}
Frequent airline passenger (4 h/week flying)	7.0×10^{-4}
Average diagnostic medical X-rays	1.4×10^{-3}
Natural background radiation	1.4×10^{-3}
Living in masonry building rather than wood	3.5×10^{-4}
Eating and Drinking	
One 12-$^1/_2$ ounce diet drink per day	7.0×10^{-4}
Average saccharin consumption	1.4×10^{-4}
Four tablespoons peanut butter per day	5.6×10^{-4}
One pint milk per day	1.4×10^{-4}
Miami or New Orleans drinking water	7.0×10^{-5}
$^1/_2$ lb charcoal broiled steak per week (cancer risk only)	2.1×10^{-5}
Alcohol, one beer per day	1.4×10^{-3}
Tobacco	
Smoker, cancer only	8.4×10^{-2}
Smoker, all effects (including heart disease)	2.1×10^{-1}
Person sharing room with smoker	7.0×10^{-3}
Air Pollution	
Polycyclic organics, all effects	1.1×10^{-3}

[a] All statistics based on U.S. data.

to human health and the environment, and therefore below regulatory concern. This label is derived from the legal principle "*de minimis non curat lex*" or "the law does not concern itself with trifles."

In a survey of Federally regulated carcinogenic chemicals, Travis et al.[26] observed regulatory patterns which indicate the existence of both a *de minimis* and a *de manifestis* level of risk. The data collected from regulating agencies indicate the following:

1. Every carcinogen with an individual risk level of 4×10^{-3} (four chances in 1000 that a chronically exposed individual may develop cancer) or above is regulated. These chemicals reside in the *de manifestis* or obvious range of risk.
2. Chemicals with individual risk levels of 1×10^{-6} or less (one chance in 1 million that a chronically exposed individual may develop cancer) are not regulated, except in one FDA decision. This uniformity in risk management decisions unofficially defines a *de minimis* level for carcinogenic chemicals.

REFERENCES

1. Chemical Carcinogens: A Review of the Science and Its Associated Principles, Office of Science and Technology Policy (OSTP), February, 1985.
2. **Rodricks, J. V.,** Risk assessment at hazardous waste disposal sites, *Hazardous Waste,* 1 (3), 333, 1984.
3. National Academy of Sciences, Risk Assessment in the Federal Government. Managing the Process, National Academy Press, Washington, D. C., 1983.
4. International Agency for Research on Cancer (IARC), Polychlorinated biphenyls and chemicals to humans, Monogr. Evaluation of the Carcinogenic Risk of Chemicals to Humans, 18, International Agency for Research on Cancer, Switzerland, 1979.
5. U. S. Environmental Protection Agency, Health Assessment Document for Tetrachloroethylene, Final Report, EPA 600/8-82/005F, Office of Health and Environmental Assessment, U.S. Environmental Protection Agency, Washington, D.C., July, 1985.
6. **Efron, E.,** *The Apocalyptics: Cancer and the Big Lie,* Simon and Schuster, New York, 1984, 322.
7. Task Force on Environmental Cancer and Heart and Lung Disease, Summary of the Workshop on Exposure to Environmental Agents, Their Metabolism and Mechanism of Toxicity, Research Needs, Project Group on Exposure and Metabolic Mechanism Report to Congress, Washington, D.C., 1981.
8. Radiological Assessment: Predicting the Transport, Bioaccumulation, and Uptake by Man of Radionuclides Released to the Environment, Report No. 76, National Council on Radiation Protection and Measurement, 1984.
9. **Miller, C. W. and Little, C. A.,** A Review of Uncertainty Estimates Associated with Models for Assessing the Impact of Breeder Reactor Radioactivity Releases, Report No. ORNL-5832, Oak Ridge National Laboratory, Oak Ridge, TN, 1982.
10. **Fox, D. G.,** Judging air quality model performance: a summary of the AMS workshop on dispersion model performance, Woods Hole, MA, September 8 to 11, *Bull. Am. Meteorol. Soc.,* 62, 599, 1980.
11. **Donigian, A. S.,** Model predictions vs. field observations: the model validation/testing process, in *Fate of Chemicals in the Environment,* Swann, R. L. and Eschenroeder, A., Eds., ACS Symp. Ser. No. 225, American Chemical Society, Washington, D.C., 1983.
12. **Wallace, L. A., Pellizzari, E. D., and Gordon, S.M.,** Organic chemicals in indoor air: a review of human exposure studies and indoor air quality studies, in *Indoor Air and Human Health,* Gammage, R. B. and Kaye, S. V., Eds., Lewis Publishers, Chelsea, MI, 1985.
13. **Crump, K. S.,** *Principles of Health Risk Assessment,* Ricci, P., Ed., Prentice Hall, Englewood Cliffs, NJ, 1985, chap. 7.
14. Site Selection Process, Phase 4A: Selection of a Preferred Site (S) Generic Risk Assessment, 2, Final Report prepared for Ontario Waste Management Corp., by the ENVIRON Corp., Washington, D.C., April 21, 1986.
15. Bureau of the Census, Statistical Abstract of the United States 1984, U.S. Government Printing Office, Washington, D.C., 1984, 104.
16. International Commission on Radiological Protection (ICRP), *Report on the Task Group on Reference Man,* Vol. 23, Pergamon Press, Oxford, 1975, 346.

17. Food and Drug Administration (FDA), Fiscal Year 1979 Total Diet Studies — Adult, NTIS: PB#: 83-112722, Food and Drug Administration, Washington, D. C., 1980.
18. U.S. Environmental Protection Agency, Water quality criteria documents, *Fed. Regist.*, 5, 79319, November 28, 1980.
19. Risk Assessment on (2,3,5-Trichlorophenoxy) Propionic Acid and 2,3,7,8-Tetrachlorodiebenzo-P-Dioxin (TCDD), Carcinogen Assessment Group, U.S. Environmental Protection Agency, Washington, D.C., EPA-600/6-81-003.
20. **Poiger, H. and Schlatter, C.,** Influence of solvents and adsorbents on dermal and intestinal adsorption of TCDD, *Food Cosmet. Toxicol.*, 18, 477, 1980.
21. **Crump, K. S.,** An improved procedure for low-dose carcinogenic risk assessment from animal data, *J. Environ. Pathol. Toxicol.*, 5(4/5), 339, 1984.
22. *Fed. Regist.*, 45, 73918, November 1980.
23. Health Effects Assessment Summary for 300 Hazardous Organic Constituents in Support of Regulatory Impact Analysis of the Land Disposal Branch (LDB) and Interim Final Incinerator Regulations of the Technical Branch (TB) or the Office of Solid Wastes, Environmental Criteria Assessment Office, U.S. Environmental Protection Agency, Washington, D.C., 1982.
24. **Crouch, E. A. C. and Wilson, R.,** *Risk/Benefit Analysis,* Ballinger, Cambridge, Mass., 1982, 165.
25. ENVIRON, Site Selection Process, Phase 4A: Selection of a Preferred Site (S) Generic Risk Assessment, 1, Final Report prepared for Ontario Waste Management Corp., by the ENVIRON Corp., Washington, D.C., December 6, 1985.
26. **Travis, C. C., Richter, S. A., Crouch, E. A. C., Wilson, R., and Klema, E.,** Cancer risk management: a review of 132 federal regulatory decisions, *Environ. Sci. Technol.*, 21, 415, 1987.

Chapter 5

HEALTH RISK ASSESSMENT FOR ORGANICS

I. INTRODUCTION

During incineration of hazardous wastes, most of the principal organic hazardous constituents (POHCs) are thermally destroyed or removed from the waste stream bottom as ash. Typically, destruction and removal efficiencies (DREs) at hazardous waste incinerators range from 99.99 to 99.9999%. However, small quantities of POHCs escape destruction and are released from the incinerator stack into the atmosphere. Even these small quantities may have deleterious impacts on human health. In this chapter, we present estimates of human inhalation exposure and health risk resulting from POHCs emitted during the incineration process.

II. THE RISK ASSESSMENT PROCESS

The purpose of a risk assessment at a hazardous waste incinerator facility is to provide estimates of human health risk associated with exposure to toxic pollutants resulting from facility operations. The risk assessment should (1) address the most significant pathways of human exposure and (2) address any specifically expressed public concerns regarding facility operations.

In the generic risk assessment that follows, we will only address direct inhalation of pollutants released from incineration stacks. Exposures resulting from ingestion of contaminated food stuffs will be addressed in Chapter 7. As stated in Chapter 4, an inhalation risk assessment at a hazardous waste incineration facility consists of the following six steps (see Figure 1):

1. Specific hazardous waste constituents are identified as the focus of the risk assessment.
2. The rate of release (micrograms per second) of these specific chemicals in incinerator stack exhaust is either measured directly or estimated from incinerator operating parameters.
3. Ground-level atmospheric concentrations of pollutants are determined within a given radius of the incineration facility using stack emission rates and atmospheric transport models.
4. These atmospheric concentrations (micrograms per meter^3) are then multiplied by the number of persons inhabiting the areas surrounding the incinerator to quantify the population exposure (person — micrograms per meter^3) and the maximum individual exposure (micrograms per meter^3).
5. The maximum individual exposure and the population exposure are multiplied by a conversion factor, which takes into account the average weight, breathing rate, and absorption of the contaminant by individuals in the population, to yield the maximum individual dose (milligrams per kilogram per day) and the population dose (person — milligrams per kilogram per day).
6. The maximum individual dose and the population dose estimate are multiplied by the chemical's unit potency factor, a measure of cancer risk resulting from a dose of 1 mg/kg of body weight per day for 70 years. Thus, the two primary estimates of cancer risk, maximum individual risk (MIR) and population risk (PR), for exposure to a chemical are obtained.

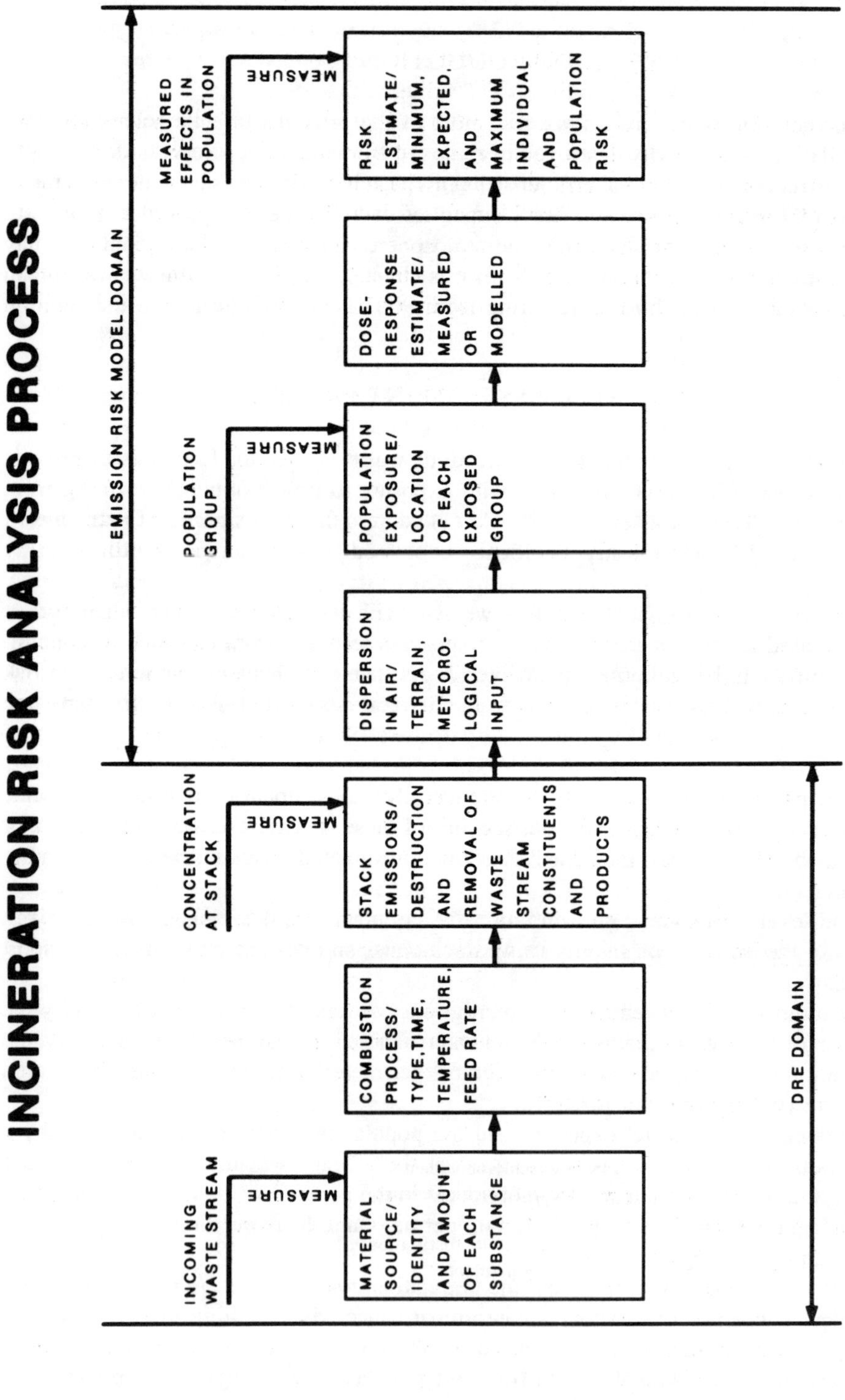

FIGURE 1. Incineration risk analysis process illustrating the six steps of an inhalation risk assessment. (Adapted from Reference 1.)

III. A GENERIC INHALATION RISK ASSESSMENT

The following risk assessment is intended to provide an overview of the human health risks resulting from POHCs emitted during the incineration process. The calculations are intended to supply only order of magnitude health risk estimates. Site-specific application of the results would require careful evaluation of the extent to which the models and parameter values used are representative of conditions prevailing at that site.

The assessment will focus only on organics in the waste stream which are released to the atmosphere during incineration. The health impact of products of incomplete combustion (PICs) and metals will be addressed in subsequent chapters.

For this assessment, three sizes of liquid injection incinerators (1, 10, and 150 $\times$ 10^6 Btu/h) are assumed to be sited in each of three different locations in the U.S. The three locations studied are a primary northern Midwest site (S-1), and two supplementary sites (S-2 and S-3) in heavily populated areas. Site S-2 is the most populous, with 8.2 million people, followed by Site S-3 with 7.4 million, and Site S-1 with 0.45 million people. The supplementary sites are included to investigate the effect of population size and distribution on human inhalation exposure and health risk. To investigate the impact of the DRE on health risk, three levels of DRE will be considered (99.0, 99.9, and 99.99%).

A. Surrogate Chemicals

A total of 237 different constituents have been identified in the more than 413 hazardous waste streams currently being incinerated in the U.S.[2] (See Chapter 1, Table 9 for a list of the twenty most prevalent constituents of incinerated waste streams.) It is beyond the scope of most risk assessments to assess the health risk associated with all the constituents in waste streams incinerated at a facility. It is therefore necessary to select a manageable number of surrogate chemicals that represent the most hazardous chemicals present in the waste streams handled at the facility. Selection of the surrogate chemicals should be based on three factors: (1) the toxicity of the chemical, (2) the quantity of the chemical handled at the facility, and (3) the potential for the chemical to be released to the environment. The list of surrogate chemicals may be different depending on whether the risk assessment is considering chronic exposure resulting from normal operating conditions or acute exposure resulting from accident conditions. Table 1 presents the list of the surrogate chemicals selected for a risk assessment at a proposed hazardous waste facility in Canada.[3]

Our generic inhalation risk assessment will focus on two waste streams: one waste stream containing carcinogenic organic chemicals and one waste stream containing toxic, but non-carcinogenic organic chemicals. The two waste streams and their constitutents are

Pesticide-related chemicals
(carcinogens)

Chloroform
Ethylene dichloride
Hexachlorobutadiene
1,1,2,2-Tetrachloroethane

Phenol/acetone distillation
chemicals
(noncarcinogens)

Phenol
Toluene
Pyridine
Phthalic Anhydride
Methyl styrene

Table 1
SURROGATE COMPOUNDS FOR INCINERATOR RISK ASSESSMENT[3]

Acrylonitrile
Benzene
Cresols
Ethanol
Formaldehyde
Hydrazine
Methyl ethyl ketone
Methylene chloride
PCB
Petroleum distillates
Phenol
Pyridine
Styrene
Tetrachloroethylene
Tetraethyl lead
Toluene
1,1,1-Trichloroethane
Trichloroethylene

B. Characterization of Stack Emissions

The rate of release (mass emissions per unit time) of specific chemicals in incinerator stack exhaust is controlled by three facility variables: (1) waste throughput, (2) chemical concentration in the waste stream, and (3) DRE. A summary of waste throughput in our generic facilities is presented in Table 2. The fractional representation of each chemical in our waste streams is considered equal to the average fractional representation of that chemical in hazardous waste streams currently being incinerated in the U.S.[4] To obtain stack emission rates, throughput is multiplied by the fractional representation of each chemical in the waste stream and by 10^{-4} to account for a 99.99% DRE (see Chapter 4, the section on stack emissions). Predicted annual stack emission rates (grams per year) for each chemical (assuming 99.99% DRE) are given in Table 3.

C. Atmospheric Dispersion

Annual-average ground-level air concentrations of representative chemical pollutants are estimated using IEM, an automated inhalation exposure methodology which employs a Gaussian-plume atmospheric dispersion model.[5] IEM is applied within a 100-km radius around each incinerator facility. The assessment areas are divided into 160 sector segments consisting of 10 concentric circles about the origin and 16 radial direction vectors (see Figure 2). Exposure and risk due to long-range transport (greater than 100 km) are not estimated.

Input parameters necessary to run the IEM atmospheric dispersion model include stack parameters for the three sizes of hazardous waste incinerators and site-specific meteorological data. Stack parameters employed are summarized in Table 4.

Region-specific meteorological data are obtained from Stability Array (STAR) data tapes[6] and from a compendium of weather statistics.[7] The remaining meteorological parameters are obtained or derived from Ruffner.[7] More detail is available in Travis et al.[8]

D. Exposure Assessment

The next step in the risk assessment process is to combine the ground-level atmospheric concentrations of the surrogate chemicals with the number of persons inhabiting the area surrounding the incinerator to quantify the population exposure (person — micrograms per

Table 2
FACILITY THROUGHPUTS

Facility	Waste (g/year)	
	Pesticide related	Phenol/acetone distillation
1 MM Btu/h	1.84E + 8	3.53E + 8
10 MM Btu/h	1.84E + 9	3.53E + 9
150 MM Btu/h	2.76E + 10	5.29E + 10

Note: MM = 1 E + 6.

Table 3
STACK EMISSION RATES OF INCINERATED CHEMICALS (99.99% DRE)

Pollutant	Fraction of waste stream (%)	Emission rate (g/year) Incinerator size × 10^6 Btu/h		
		1	10	150
Pesticide-related waste				
Chloroform	6.1	1.12E + 3	1.12E + 4	1.68E + 5
Ethylene dichloride	19.0	3.50E + 3	3.50E + 4	5.23E + 5
Hexachlorobutadiene	9.2	1.69E + 3	1.69E + 4	2.54E + 5
1,1,2,2-Tetrachloroethane	2.4	4.42E + 2	4.42E + 3	6.62E + 4
Phenol-acetone distillation waste				
Phenol	0.003	6.87E − 1	6.87E + 0	1.03E + 2
Toluene	0.6	3.91E + 3	3.91E + 4	5.87E + 5
Pyridine	1.3	2.68E + 2	2.68E + 3	4.01E + 4
Phthalic anhydride	0.001	2.05E − 1	2.05E + 0	3.09E + 1
Methyl styrene	1.8	3.72E + 3	3.72E + 4	5.55E + 5

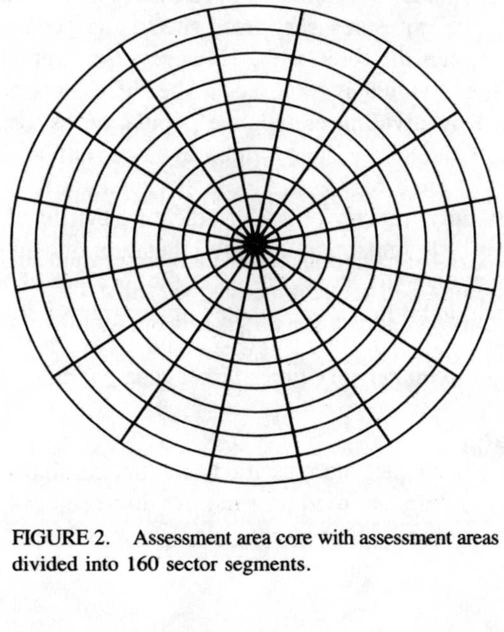

FIGURE 2. Assessment area core with assessment areas divided into 160 sector segments.

Table 4
STACK PARAMETERS EMPLOYED IN ATMOSPHERIC DISPERSION MODELING

	Liquid injector size ($\times$ 10^6 Btu/h)		
Parameter	**1**	**10**	**150**
Height (m)	15.24	30.48	30.48
Exit gas temperature (K)	355	355	355
Exit gas velocity (m/sec)	1.10	11.0	41.1
Diameter (m)	0.61	0.61	1.22
Generic gas emission rate (g/sec)	1.0	1.0	1.0

Table 5
TOTAL COLLECTIVE EXPOSURE FOR AN LI-10 INCINERATOR (99.99% DRE) AT THE S-1 SITE

Pollutant	Total value (person-μg/m^3)
Chloroform	7.13E + 0
Ethylene dichloride	1.02E + 1
Hexachlorobutadiene	1.61E + 0
1,1,2,2-Tetrachloroethane	4.96E − 1
Phenol	6.61E − 4
Toluene	6.30E + 0
Pyridine	3.93E − 1
Phthalic anhydride	1.95E − 4
Methyl styrene	3.90E + 0

meter^3) and the maximum individual exposure (micrograms per meter^3). Table 5 lists estimates of population exposure (person — micrograms per meter^3) at the primary site (S-1) for a 10 $\times$ 10^6 Btu/h liquid injection incinerator performing at 99.99% DRE.

To determine the effect of population size and distribution on inhalation exposure, results from the three sites are compared. Table 6 lists expected cumulative population exposures (person — micrograms per meter^3) to chloroform released from an LI-10 incinerator at each site. Cumulative population exposures vary from site to site because total population size and distribution vary between the three sites. However, the average air concentrations of chloroform to which an individual is exposed at the three sites are very similar. These numbers can be obtained by dividing cumulative population exposures (Table 6) by total population size at each of the sites. The resulting average individual inhalation intakes of chloroform at the three sites are 8.65E − 2, 1.95E − 1, and 9.20E −2 μg/yd, respectively. These numbers differ by only a factor of 2.25, indicating that local meteorology and population distributions have little impact on average inhalation intakes.

The population distribution will, however, affect the location of the maximally exposed individual. Maximum individual inhalation intakes (micrograms per yard) to chloroform are estimated to be 6.33E − 1 (S-1), 1.10E + 0 (S-2), and 1.65E + 0 (S-3). These maximum individual intakes are approximately six times larger than average individual intakes.

E. Health Risk Estimates

Chemicals studied in this chapter are classified as either carcinogens or noncarcinogens, and separate measures of toxicity are used for each. As discussed in Chapter 4, carcinogenic toxicity is estimated in terms of lifetime excess cancer risk, while noncarcinogenic toxicity

Table 6
CUMULATIVE POPULATION EXPOSURE TO CHLOROFORM AT THE THREE INCINERATION SITES

Distance (km)	S-1	S-2	S-3
0.2	0	0	1.74E − 1
0.4	0	0	6.10E − 1
0.6	0	6.61E − 1	9.36E − 1
1.0	0	3.00E + 0	1.20E + 0
2.0	2.05E − 1	3.11E + 0	1.87E + 0
3.0	5.72E − 1	5.35E + 1	2.33E + 0
4.0	1.02E + 0	6.70E + 1	3.08E + 0
5.0	1.41E + 0	7.42E + 1	4.61E + 0
6.0	1.75E + 0	1.06E + 2	7.52E + 0
10.0	1.82E + 0	1.28E + 2	1.36E + 1
20.0	4.99E + 0	1.92E + 2	3.57E + 1
40.0	5.73E + 0	2.70E + 2	9.74E + 1
60.0	6.55E + 0	2.90E + 2	1.11E + 2
80.0	6.90E + 0	2.92E + 2	1.18E + 2
100.0	7.12E + 0	2.96E + 2	1.23E + 2

Note: Exposure given in person-μg/m³.

Table 7
HEALTH RISK ESTIMATORS FOR CHEMICALS IN GENERIC WASTES

Pollutant	Health risk esimator	
	Excess risk[a] (mg/kg/d)$^{-1}$	ADI[b] (mg/d)
Chloroform	1.8E − 1	NA[c]
Ethylene dichloride	3.7E − 2	NA
Hexachlorobutadiene	7.75E − 2	NA
1,1,2,2-Tetrachloroethane	2.0E − 1	NA
Phenol	NA	6.8E + 0
Toluene	NA	1.3E + 2
Pyridene	NA	5.4E + 0
Phthalic anhydride	NA	2.1E + 0
Methyl styrene	NA	1.7E + 1

[a] Excess risk of developing cancer after a lifetime exposure.[10]
[b] ADI = acceptable daily intake.[10]
[c] NA = not applicable.

is measured in terms of an acceptable daily intake (ADI). The estimate of excess lifetime human cancer risk is used for all compounds in the pesticide-related waste. The ADI value is used for all compounds in the phenol/acetone distillation waste.

Toxicity data are taken from a health effects assessment summary for 300 hazardous organic constituents prepared by the EPA's Environmental Criteria Assessment Office (ECAO).[10] Health effects data for compounds considered in the present inhalation risk assessment are summarized in Table 7.

1. Carcinogenic Risk

Potency factors for chemical carcinogens (derived by the EPA from available human

Table 8
MAXIMUM INDIVIDUAL RISK FROM INCINERATION OF PESTICIDE-RELATED WASTE (99.99% DRE)

Site	LI-1	LI-10	LI-150
S-1	2.8E − 8	7.8E − 8	1.2E − 7
S-2	6.5E − 8	1.8E − 7	2.8E − 7
S-3	7.5E − 7	2.1E − 6	3.2E − 6

Table 9
EXPECTED POPULATION RISK FROM INCINERATION OF PESTICIDE-RELATED WASTE AT SITE S-1

DRE (%)	LI-1	LI-10	LI-150
99.99	1.9E − 6	5.7E − 6	2.3E − 5
99.90	3.7E − 6	2.0E − 6	1.7E − 4
99.00	2.3E − 6	1.6E − 4	1.7E − 3

Note: Risk expressed in cancers per year. Population, 0.45×10^6.

epidemiological data and from animal studies) are purposefully conservative. As discussed in Chapter 4, the excess risk represents the upper 95% confidence limit of the largest possible slope of a linear dose-response curve consistent with empirical human or animal data. Thus, the true or expected risk associated with a chemical is smaller than this maximum risk with 95% confidence.

Estimates of excess lifetime cancer risk to the maximally exposed individual from incineration of pesticide-related waste are presented in Table 8. Excess cancer risk is less than 3.2×10^{-6} for all incinerator sizes at all sites. Estimates of MIR at the different sites differ by about a factor of 30, depending on the proximity of the maximally exposed individual to the incineration site.

Estimates of the PR over 70 years associated with incinerating the generic pesticide-related waste stream are given in Table 9. These estimates are the sum of the cancers caused by all four of the carcinogenic chemicals in the pesticide-related waste. At 99.99% DRE, they indicate that the probability of having one cancer death per year in the entire population at site S-1 is roughly six in a million for an LI-10 incinerator. In other words, if the generic pesticide-related waste were incinerated at site S-1 continuously over a period of one million years, one might expect six additional cancer deaths in the population of 0.45 million people.

2. *Noncarcinogenic Risk*

None of our four selected constituents of the phenol/acetone distillation wastes are known human carcinogens. We therefore use the ADI as a measure of noncarcinogenic risk. Estimates of average daily intake of these constituents (as a measured fraction of the ADI) are listed in Table 10. As can be seen, the estimated daily intake of these organics are a small fraction of the ADI.

Table 10
AVERAGE DAILY INTAKE FROM PHENOL/ACETONE DISTILLATION WASTE (PRESENTED AS A FRACTION OF ADI)

		Incineration facility (99.99% DRE)		
Pollutant	**ADI (mg/d)**	**LI-1**	**LI-10**	**LI-150**
Phenol	6.8	5.8E − 13	2.5E − 12	1.8E − 11
Toluene	134	2.8E − 10	1.2E − 9	8.5E − 9
Pyridine	5.4	4.1E − 10	1.9E − 9	1.4E − 8
Phthalic anhydride	2.1	3.2E − 13	2.4E − 12	2.5E − 11
Methyl styrene	170	9.3E − 11	5.8E − 10	5.7E − 9

IV. DISCUSSION

Both carcinogenic and noncarcinogenic human health risks from POHCs released during incineration are small. For incineration of pesticide-related waste at a site with a population of 0.45×10^6, the expected annual number of cancers is less than 5.7×10^{-6} for an LI-10 incinerator (see Table 9). The excess cancer risk over 70 years to the maximally exposed individual at site S-1 from incineration of pesticide-related waste is less than 1.2×10^{-7} for a DRE of 99.99% (see Table 8). Investigation of incinerators with populations residing closer to the site indicates that excess cancer risk to the maximally exposed individual could be increased by a factor of about 30.

In general, the maximum individual inhalation risk from carcinogenic POHCs released during incineration appears to be less than 1×10^{-6}. This is an acceptable level of risk by most standards. The health impact of PICs and metals will be addressed in subsequent chapters.

REFERENCES

1. PEER Consultants, Inc., Supporting Documentation for the RCRA Incinerator Regulation 40 CFR 264 Subpart O-Incinerators, Office of Solid Wastes, U.S. Environmental Protection Agency, Washington, D.C., October, 1985.
2. MITRE Corporation, Composition of Hazardous Waste Streams Currently Incinerated, Office of Solid Wastes, U.S. Environmental Protection Agency, Washington, D.C., April, 1983.
3. Site Selection Process. Phase 4A: Selection of Preferred Site (S) Generic Risk Assessment, 2, Final Report, prepared for Ontario Waste Management Corp. by the ENVIRON Corp., Washington, D.C., April 21, 1986.
4. Background Document, Resource and Recovery Act, Subtitle C, Identification and Listing of Hazardous Waste, Office of Solid Wastes, U.S. Environmental Protection Agency, Washington, D.C., 1980.
5. **O'Donnell, F. R. and Holton, G. A.,** An Automated Methodology (IEM) for Assessing Inhalation Exposure to Hazardous Waste Incineration Emissions, in Incineration and Treatment of Hazardous Waste, Proc. 9th Annu. Res. Symp. Fort Mitchell, Kentucky, May 2 to 4, 1983, U.S. Environmental Protection Agency, Cincinnati, OH, 1984.
6. National Oceanic and Atmospheric Administration, Seasonal and Annual Distribution by Pasquill Stability Classes STAR Program, National Climatic Center, U.S. Department of Commerce, Asheville, NC, 1974.
7. **Ruffner, J. A.,** *Climates of the States,* Vols. 1 and 2, Gayle Research, Book Tower, Detroit, MI, 1978
8. **Travis, C. C., Etnier, E. L., Holton, G. A., O'Donnell, F. R., Hetrick, D. M., Dixon, E., and Harrington, E. S.,** Inhalation Pathway Risk Assessment of Hazardous Waste Incineration Facilities, ORNL/TM-9096, Oak Ridge National Laboratory, Oak Ridge, TN, October, 1984.

9. Health Effects Assessment Summary for 300 Hazardous Organic Constituents in Support of Regulatory Impact Analysis of the Land Disposal Branch (LDBO) and Interim Final Incinerator Regulations of the Technical Branch (TB) of the Office of Solid Wastes, Environmental Criteria Assessment Office, U.S. Environmental Protection Agency, Cincinnati, OH, 1982.
10. Summary of Current Acceptable Daily Intakes for Various Pollutants, Environmental Criteria Assessment Office, U.S. Environmental Protection Agency, Cincinnati, OH, 1983.

Chapter 6

HEALTH RISK ASSESSMENT FOR METALS

I. INTRODUCTION

The focus of current federal regulations controlling hazardous waste incineration is in the destruction and removal of the principal organic hazardous constituents (POHCs). The control of trace metal emissions is not explicitly addressed in these regulations. This raises a potential problem since, unlike organics, metals in hazardous waste streams are not thermally destroyed during incineration. Instead, they are partially or completely vaporized during the combustion process. After leaving the combustion zone, the varporized metals cool and either condense to form particles or condense on the surface of particles already present in the combustion gas. Air pollution control devices (APCDs) at the incinerator will remove some fraction of these metal-containing particulates, but significant fractions may be released to the atmosphere. In the present chapter, we estimate possible human health risk resulting from inhalation exposure to metals released during hazardous waste incineration.

The approach taken in the present chapter is a probabilistic one. The lack of empirical data on metal concentrations in waste streams limits our investigation to three metals: lead (Pb), cadmium (Cd), and chromium (Cr). First, we determine probability distributions for concentrations of these three metals in hazardous waste streams. We then determine probability distributions for individual exposures which account for variation in the concentrations of these metals in waste streams, differences in facility design across the U.S., and differences in population distributions surrounding incinerators. Thus, our analysis provides estimates of confidence intervals which reflect the variability of the potential metal exposures resulting from hazardous waste incineration throughout the U.S. Estimates of carcinogenic and noncarcinogenic health risks associated with such exposures are calculated, and quantitative estimates of variation in these risks are presented. For comparison, we also estimate human inhalation exposures and health risks resulting from measured metals emissions rates at several of the trial burn facilities. Our conclusions are markedly different from those of previous studies[8,10] and suggest the need for more stringent measures to minimize metal emissions from hazardous waste incineration.

II. THE METALS DATA BASE

Data on metal concentrations in incinerated hazardous waste streams are scarce and usually based on estimates rather than precisely measured values. The best data available come from an Environmental Protection Agency (EPA) compilation of metal concentrations in 71 waste streams,[3] data gathered at EPA trial burns,[12] and an EPA survey of 150 hazardous waste streams generated by industry,[16] but not necessarily incinerated. The following sections summarize the available metals data for waste stream concentrations, incinerator destruction and removal efficiencies (DREs), and stack emissions from incinerated waste.

A. Metal Concentration in Waste Streams

Table 1 computes the range of 32 metals concentrations found in the 71 waste streams reviewed by Haus and Gerstein.[3] It is important to note that most of these data are based on very limited sampling and analysis of the waste streams. As can be seen, six metals (arsenic, iron, Pb, magnesium, potassium, and sodium) are reported to have been present in some waste streams at concentrations greater than 100,000 μg/g. At this concentration, the metal comprises over 10% of the waste stream by weight. Five metals (aluminum, Cd,

Table 1
RANGE OF TRACE METAL CONCENTRATIONS IN SELECTED HAZARDOUS WASTES (μg/g)[a]

Trace metals	Range of conc
Aluminum (Al)	<1—14,000
Antimony (Sb)	0.1—276
Arsenic (As)	0.0—100,000
Barium (Ba)	0.01—8,000
Beryllium (Be)	0.001—7
Cadmium (Cd)	0.0—10,400
Chromium (Cr)	0.0—20,000
Cobalt (Co)	0.01—83
Copper (Cu)	0.0—45,300
Hafnium (Hf)	NR
Indium (In)	NR
Iron (Fe)	<0.1—253,000
Lead (Pb)	0.0—130,000
Magnesium (Mg)	0.001—300,000
Manganese (Mn)	0.02—250
Mercury (Hg)	0.0—220
Molybdenum (Mo)	0.02—120
Nickel (Ni)	0.0—6,550
Platinum (Pt)	4[c]—6[c]
Potassium (K)	1.4—100,000
Selenium (Se)	0.005—52
Silver (Ag)	0.0007—20
Sodium (Na)	ND—110,000
Tantalum (Ta)	NR
Tellurium (Te)	1.9—2.6
Tin (Sn)	0.01—140
Titanium (Ti)	<0.01—540
Tungsten (W)	0.1
Vanadium (V)	<0.01—3,500
Yttrium (Y)	0.01—0.6
Zinc (Zn)	0.1—17,500
Zirconium (Zr)	0.03—6,000

Note: NR = not reported; ND = not detected; heat content (kJ/kg) range = 470 to 45,200.

[a] Includes trace metal concentrations of all hazardous wastes reviewed except sewage sludge; not all trace metals were analyzed at all sites; data are based on 67 different waste streams at over 100 sites, plus 319 samples of three waste oil streams from an unknown number of sites.
[b] Moisture-free value.
[c] Data are reported to be suspect.

From Haus, S. and Gerstein, R., working paper prepared by the MITRE Corporation for the Office of Solid Waste, U.S. Environmental Protection Agency, Washington, D.C., November 1983.

Table 2
RANGE OF METAL CONCENTRATIONS IN WASTE FEEDS (μg/g)

	Ross	Plant B	American Cyanamid	DuPont	TWI
Silver	<0.03—<0.3	<1	<0.45	<2.6	<1
Arsenic	<3.5—<15	<20	11	<24.0—45.2	<23
Barium	0.83—1460	5.3—220	<0.24	<3.6—326.0	16—1200
Beryllium	<0.15—<0.3	<1	<0.5	<1.8	<0.3—1.3
Cadmium	<0.016—224	<1—7.8	<0.84	<5.0—175	<2—64
Chromium	<0.034—542	2.6—83	250	<5.2—2170.0	8.3—320
Mercury	<3—<4	<10	4	<22.0	<50
Nickel	<0.063—3100	1.9—10.3	170	<61.0—6670.0	<4—18
Lead	<0.25—2090	<10—200	1	<19.0—9830.0	<23—1500
Antimony	<0.184—437	<10—87	<2	<12.0—35.2	<24
Selenium	<2—<21	<100—340	10	<470—546.0	<90—<290
Thallium	<4—<9	<20	5	<23.0	<22

From Trenholm, A., Gorman, P., and Jungclaus, G., report prepared by Midwest Research Institute (MRI) for the U.S. Environmental Protection Agency, Cincinnati, OH, 1984.

Cr, copper, and zinc) were present at concentrations between 10,000 and 100,000 μg/g. Of these 11 metals, four (arsenic, Pb, Cd, and Cr) are suspected to be human carcinogens.

Trenholm et al.[12] measured metal concentrations in waste feeds at the EPA trial burn facilities. Table 2 presents a summary of the range of metal concentrations in the waste feeds actually incinerated during the trial burns. Notice the barium, Cr, nickel, and Pb were present at concentrations greater than 1000 μg/g. However, comparison of the Trenholm et al. data with the metal concentrations reported by Haus and Gerstein[3] indicates that the metal content of waste streams incinerated during the trial burns was not fully representative of the metal content of waste streams throughout the U.S. Of the 12 metals measured by Trenholm et al.,[12] 11 are included in the Haus and Gerstein[3] study. In most cases, the values from Trenholm et al.[12] (Table 2) are in the lower part of the range reported by Haus and Gerstein;[3] the highest arsenic level in the trial burn study is four orders of magnitude lower than the highest arsenic level in the Haus and Gerstein study. Only two metals, antimony and selenium, were found at higher levels by Trenholm et al.[12] These comparisons indicate that the toxic metal content of waste streams incinerated at the EPA trial burns was lower than found in the Haus and Gerstein[3] survey.

To assess the variability of metal concentrations in hazardous waste streams, we reviewed data from an EPA survey of 150 hazardous waste streams generated by industry, but not necessarily incinerated.[16] The most frequently identified metals were Pb (41 waste streams), Cr (33 waste streams), and Cd (15 waste streams). These are the only metals for which there are enough data to develop probability distributions. Figure 1 displays the concentration data for these metals on probability paper to determine whether the measured concentrations are lognormally distributed. (This is a commonly used technique for determining the appropriateness of a particular probability distribution to describe a set of data.[1,9]) Since a straight line reasonably fits the measured concentration data for Pb and Cr, the assumption of a lognormal distribution is determined to be appropriate. Due to the limited data set available for Cd (only 15 observations), the fit to a lognormal distribution is not as good as for the other two metals. However, since we did so for Pb, and Cr, we will assume that the concentrations of Cd in hazardous waste streams throughout the U.S. are lognormally distributed.

The range of Pb, Cd, and Cr concentrations reported by Haus and Gerstein[3] (see Table 1) compare very well with the distributions in Figure 1. The measured values from the Haus

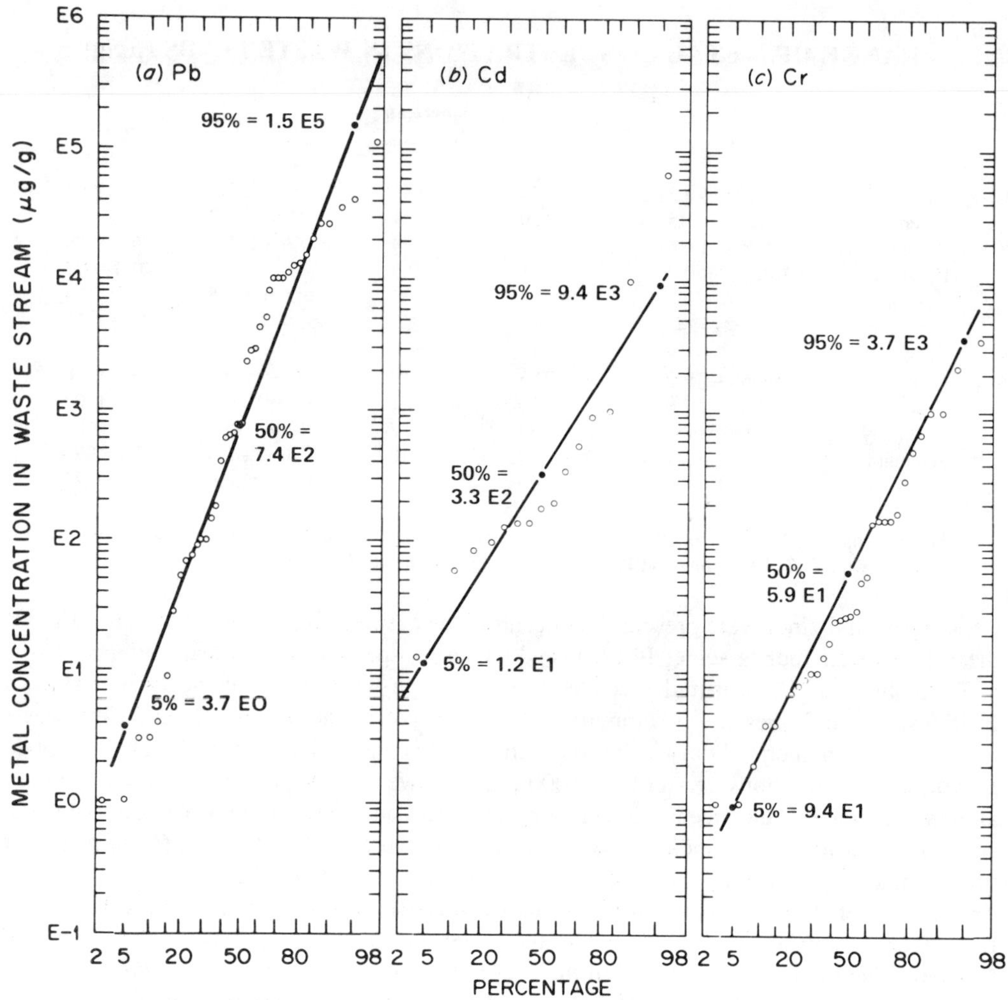

FIGURE 1. Probability distribution for the concentrations (micrograms per gram) of lead (a), cadmium (b), and chromium (c) for waste streams in which they were present.

and Gerstein[3] study reach approximately the 95th percentile for Pb and Cd and are just off the scale for Cr. The range of Pb, Cd, and Cr concentrations gathered at the EPA trial burns[12] is considerably smaller than the range found in Figure 1, reaching the 42nd percentile for Cd, the 78th percentile for Pb, and the 92nd percentile for Cr. Thus, the data indicate that the waste streams in the EPA trial burns were lower in metal content than one would expect from the distributions based on the Vogel et al.[16] data set.

B. Metal Emission Factors

Table 3 compiles metal emission factors for six metals analyzed at four EPA trial burn facilities. An emission factor is the percentage of a metal in the waste stream that is emitted to the atmosphere during incineration. Emission factors for heavy metals range as high as 51%. In general, however, the emission factor for metals at incinerators with APCDs is around 3%, with roughly half of the remainder of heavy metals going into the scrubber water of the APCD, and the other half going into bottom ash.[12]

Based on trial burn data, the geometric means of the emission factors for Pb, Cd, and Cr are 10.2, 6.4 and 1.8%, respectively. Thus, the emission factor for Pb is about 3.4 times that of an average metal.

Table 3
METAL EMISSIONS AS A PERCENTAGE OF METALS INPUT AT FOUR TRIAL BURN FACILITIES

Site	APCD	Run no.	Sb	Ba	Cd	Cr	Pb	Ni
Ross	Yes	1	[a]	0.16	0.96	0.88	8.4	0.42
		2	5.3	0.13	1.1	0.73	9.7	0.05
		3	9.1	0.13	12.3	1.2	11.4	0.26
American Cyanamid	No	3	[a,b]	[a,b]	[a,b]	34	[b]	47
DuPont	Yes	1	[b]	1.6	3.1	0.43	4.2	0.28
		2	[b]	13.3	0.94	[b]	1.2	0.091
		3	[b]	0.75	2.4	[b]	1.8	0.25
TWI	Yes	1	[a,b]	0.24	26	1.8	25	[b]
		2	[a,b]	0.26	50	1.9	51	[b]
		3	[a,b]	0.15	25	1.0	26	[b]
		4	[a,b]	0.24	37	1.8	43	[b]
Range			**5.3—9.1**	**0.1—13**	**1—50**	**0.4—2**	**1.2—51**	**0.01—47**

[a] Amounts in waste inputs contained < values.
[b] Amounts in particulate samples contained < values.

From Trenholm, A., Gorman P., and Jungclaus, G., report prepared by Midwest Research Institute (MRI) for the U.S. Environmental Protection Agency, Cincinnati, OH, 1984.

Table 4 presents the percentage of input metals contained in both controlled and uncontrolled emissions of metals from hazardous waste incinerators.[3] Controlled emissions refer to those emissions remaining in the stack gas after passage through the APCD. Uncontrolled emissions are those releases present in the combustion gas of incinerators without air pollution control equipment. Comparison of the Trenholm et al.[12] data with the Haus and Gerstein[3] data indicate similar patterns.

C. Metal Emission Rates

Trenholm et al.[12] measured metal emission rates during the EPA trial burns. Table 5 presents a summary of these data. Notice that Pb, Cr, and Cd were present in the particulate emissions at concentrations as high as 100,000; 47,500; and 4300 μg/g, respectively. At these concentrations, Pb and Cr comprise 10 and 4.8% of the particulate mass by weight. Comparison of Table 5 with Table 2 indicates the concentrations of Pb, Cr, and Cd in the particulate emissions were much higher than their concentrations in the waste feeds that were incinerated. This phenomenon is known as enrichment. The enrichment factor is defined as the ratio of the concentration of a trace metal in particulates to the concentrations of that trace metal in the input waste. Enrichment occurs because more metals condense on the surface of the finer particulates, and smaller particles are not removed efficiently from the emission gases by air pollution control equipment.

Table 6 presents measurement of metal emissions from an operating incinerator in the Chicago area.[4,5] These data indicate higher emission rates for metals than observed at the EPA trial burns. For example, the largest measured Pb emission rate at the Chicago incinerator is 8.2 times larger than any Pb emission rate measured during the trial burns.

The rate of release (mass per unit time) of metals in incinerator stack emissions is controlled by three variables: metal concentration in the waste feed, waste throughput, and DRE. We

Table 4
PERCENTAGE OF INPUT METALS CONTAINED IN CONTROLLED AND UNCONTROLLED EMISSIONS FROM HAZARDOUS WASTE INCINERATORS

Metal	Controlled emissions[a]	Uncontrolled emissions[b]
Antimony (Sb)	>7.5	>14
Barium (Ba)	0.1—5.2	6.3
Beryllium (Be)	<2.3	—
Cadmium (Cd)	2.1—31	>63
Chromium (Cr)	0.4—14	56
Lead (Pb)	2.4—32	>62
Mercury (Hg)	>1.3	—
Nickel (Ni)	0.2—8	—
Selenium (Se)	>3.6	>56
Silver (Ag)	>4.3	—

[a] Scrubber method of air pollution control used; number of sites sampled = 4; and not all trace metals were sampled at all four sites.
[b] Number of sites sampled = 1.

From Haus, S. and Gerstein, R., working paper prepared by the MITRE Corporation for the Office of Solid Waste, U.S. Environmental Protection Agency, Washington, D.C., November 1983.

Table 5
RANGE OF METAL EMISSIONS FOR FIVE HAZARDOUS WASTE INCINERATOR TESTS

	Emission range	
Metals	μg/g of particulate	g/min
Antimony	300—15,200	<0.012—1.2
Arsenic	900[a]	0.019[a]
Barium	41—3,090	0.0032—0.12
Beryllium	<0.32—6.2	<0.0001—0.002
Cadmium	140—4,300	0.0033—0.087
Chromium	668—47,500	0.0066—0.47
Lead	3,100—100,000	<0.016—7.3
Mercury	56[a]	0.0043[a]
Nickel	200—49,000	<0.004—0.57
Selenium	<500—61,600	<0.046—4.5
Silver	7.6—1.880	0.001—0.0064
Thallium	<120—150	<0.004—0.012

Note: < indicates that the concentration of metal was below the detection limit.

[a] Detected during only one run; below detection limits in all other runs.

Adapted from Trenholm, A., Gorman, P., and Jungclaus, G., report prepared by the Midwest Research Institute (MRI) for the U.S. Environmental Protection Agency, Cincinnati, OH, 1984.

Table 6
METAL EMISSION RATES AT A CHICAGO INCINERATOR

Metals	μg/g of particulate	g/min
Ag	<0.95—<16	0.002—<0.03
Al	63—6,300	0.12—12
As	ND	ND
Be	<0.63—6.3	0.001—0.01
Cd	3.2—16	0.006—0.03
Co	13.0—95	0.025—0.18
Cr	95—320	0.18—0.61
Cu	63—320	0.12—0.61
Fe	1,300—6,300	2.47—11.97
Mg	32—280	0.061—0.53
Mo	9.5—63	0.018—0.12
Ni	32—250	0.061—0.48
Pb	6,310—31,500	12—59.9
Sb	16—63	0.03—0.12
Sn	160—950	0.30—1.81
Ti	23—320	0.044—0.61
Tl	<32—63	0.061—0.12
Zn	1,100—5,200	2.09—9.9

Note: ND = not detected.

will assume a waste throughput of 1×10^9 g/year for a 10×10^6 Btu/h incinerator, and a DRE of 89.8% for Pb, 93.6% for Cd, and 98.2% for Cr. Based on these assumptions and the probability distributions for concentrations of Pb, Cd, and Cr in waste streams, as given in Figure 1, predicted distributions for annual stack emission rates (grams per year) for these three metals are given in Figure 2.

The range of Pb, Cd, and Cr emission rates reported at the EPA trial burns (see Table 5) compare well with the distributions in Figure 2. The measured values from the trial burns reach the 92nd percentile for Pb and the 98th percentile for Cr. Measurements of metal emission rates at an operating incinerator in the Chicago area[4] (Table 6) further support the reasonableness of the distributions we have predicted for Pb, Cd, and Cr. The range of Cd emissions lies just below the midrange of that distribution (20 to 50th percentiles), while Cr emission rates correspond to the 92nd to 97th percentile. Pb emission rates at the Chicago incinerator were above the 95th percentile to off the scale.

III. PROBABILISTIC RISK ASSESSMENT

The following summarizes possible inhalation risk estimates resulting from exposure to Pb, Cd, and Cr emissions at hazardous waste incineration facilities. The approach taken is a probabilistic one. The end result of our analysis will be an estimate of the probability of exceeding certain health standards for Pb, Cd, and Cr at an arbitrary incinerator in the U.S.

A. Human Exposures

The maximum and average individual exposures (micrograms per meter^3) resulting from a unit (1 g/sec) release of a pollutant have been determined for 31 actual incinerators located in various parts of the U.S.[6,8] The maximum and average individual exposures (micrograms per meter^3) were determined within a 50 km radius of the incinerators using a slightly modified version of the Industrial Source Complex Long-Term Model (ISCLTM), which is

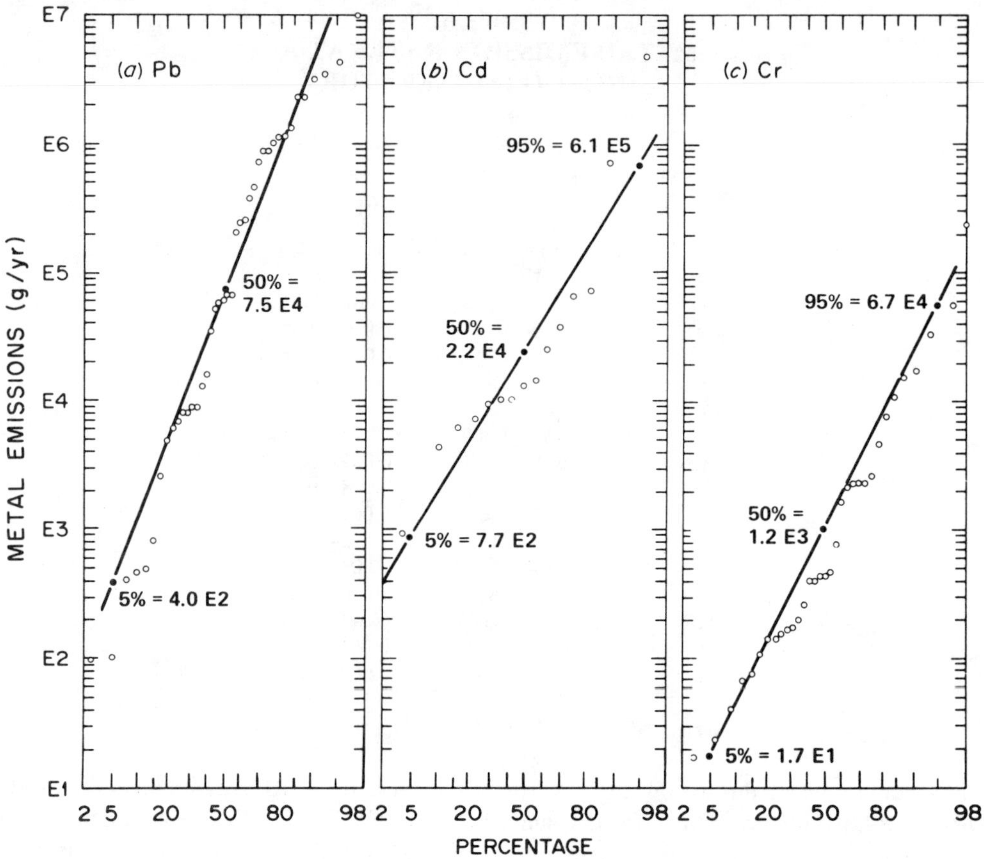

FIGURE 2. Probability distributions for the emissions of lead (a), cadmium (b), and chromium (c), grams per year.

a Gaussian-plume model developed for the EPA.[2,7] Figure 3 displays the data on probability paper to determine whether the predicted exposures are lognormally distributed. Since a straight line reasonably fits the data, the assumption of a lognormal distribution is determined to be appropriate.

The probability distributions of average individual and maximum individual exposures given in Figure 3 are for an emission rate of 1 g/sec. These distributions reflect the variability of average individual and maximum individual exposures (micrograms per meter^3) resulting from incineration of hazardous waste at an arbitrary incinerator site in the U.S. They take into account differences in incinerator design (e.g., stack height, stack gas temperatures, and stack exit velocities), differences in meteorological conditions at various incinerators, and differences in population distributions around incinerator sites. For the incinerators considered, stack heights varied from 0.61 m to 64 m, stack gas exit temperatures varied from 14 to 871°C, and stack exit velocities varied from 1 to 45 m/sec. Population sizes around the incinerators varied from 12,700 to 6,110,000. Figure 3 indicates that average and maximum individual exposures (micrograms per meter^3) resulting from a unit (1 g/sec) release rate are lognormally distributed with a mean value of 1.95E − 2 and 6.64E − 2 $\mu g/m^3$, respectively.

B. Health Effects

Metals studied in the present chapter are classified as either carcinogens or noncarcinogens. Carcinogenic toxicity is estimated in terms of lifetime excess cancer risk, while noncarcin-

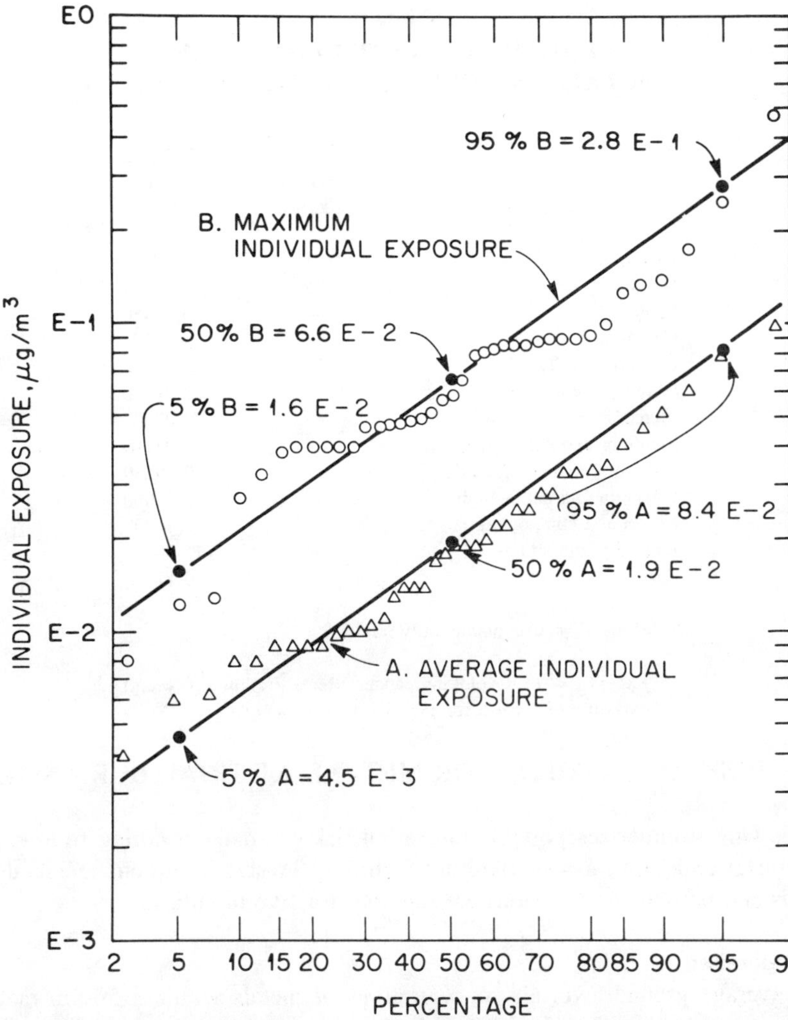

FIGURE 3. Probability distributions of average individual exposure (a) and maximum individual exposure (b), micrograms per cubic meter, for an emission rate of 1 g/sec.

ogenic toxicity is reflected in acceptable daily intake (ADI). Health risk estimators for metals considered in this chapter are summarized in Table 7. The EPA does not use ADIs for carcinogenic substances (arsenic, beryllium, Cd, inhaled Cr, and nickel), but they have been included here because of the other types of toxicity these metals may exhibit.

Figure 4 presents probability distributions for daily intakes (milligrams per day) of Pb, Cd, and Cr resulting from incineration of metals at an arbitrary incinerator in the U.S. These distributions were obtained by combining the distributions for metal emission rates (Figure 2) with the distribution for average individual exposure (Figure 3) and multiplying by a conversion factor to account for breathing rate. The ADIs are shown as are daily intakes corresponding to the 10^{-6} lifetime cancer risk level for Cd and Cr.

The "A" distribution in Figure 4 is based on emissions from a 10×10^6 Btu/h incinerator. The "B" distributions in those figures are for a 150×10^6 Btu/h incinerator, and the "C" distributions are for the worst case scenario (maximally exposed individual in the vicinity of a 150×10^6 Btu/h incinerator). For the worst case scenario, the DREs for Pb and Cd are reduced to 80 and 65%, respectively, to correspond to the lowest values measured by Trenholm et al.[12]

Table 7
HEALTH RISK ESTIMATORS FOR METALS IN GENERIC WASTE STREAMS

	Health risk estimator	
Pollutant	Excess risk[a] $(mg/kg/d)^{-1}$	ADI (mg/d)
Arsenic	50.1	0.07
Beryllium	2.6	0.00071
Lead	—	0.00450
Cadmium	7.8	0.0179
Chromium (VI)	41	0.175
Silver and compounds	—	0.01600
Mercury	—	0.0200
Thallium and compounds	—	0.03730
Antimony and compounds	—	0.29200
Selenium and compounds	—	0.70000
Nickel and compounds	1.15	1.4600
Zinc and compounds	—	15.0000
Barium	—	0.179

Note: ADI = acceptable daily intake.[14]

[a] Excess risk of developing cancer after a lifetime (70 years) exposure to 1 mg/d for each kg of body weight.[15]

IV. RISK ASSESSMENT FOR METALS AT TRIAL BURN SITES

The following summarizes possible inhalation risk estimates resulting from exposure to measured metal emission rates at trial burn facilities. Twelve hazardous metals detected in waste feeds and particulate emissions are reported for five facilities.

A. Air Dispersion

Annual-average ground-level air concentrations of metals within a 50 km radius of incinerators are estimated using the ISCLTM. The mean value for the maximum individual exposure resulting from a 1 g/sec release is 6.6E − 2 $\mu g/m^3$, while the mean value for the average individual exposure is 1.9E − 2 $\mu g/m^3$ (see Figure 3).

When the stack emission rates of the hazardous metals given in Table 5 are combined with mean individual exposures resulting from 1 g/sec releases (Figure 3), ranges are obtained for individual exposures resulting from incineration at the five facilities where metal releases were monitored. These ranges for minimum and maximum individual exposures are presented in Table 8.

B. Health Risk Estimates

Both carcinogenic and noncarcinogenic health risk are estimated using the risk estimators and procedures described previously for Pb, Cd, and Cr (see the section on health effects).

1. Carcinogenic Risk

Assuming an average exposed individual receives 0.0195 $\mu g/m^3$ for a unit release of 1 g/sec (the mean of distribution A in Figure 3), lifetime risk values are calculated for arsenic, Cr, beryllium, Cd, and nickel (Table 9). For the rate of release at one incinerator at which a measurable amount of arsenic was emitted, the estimate of carcinogenic risk is 1.4E − 8. For the range of beryllium emissions measured, the maximum estimate of carcinogenic

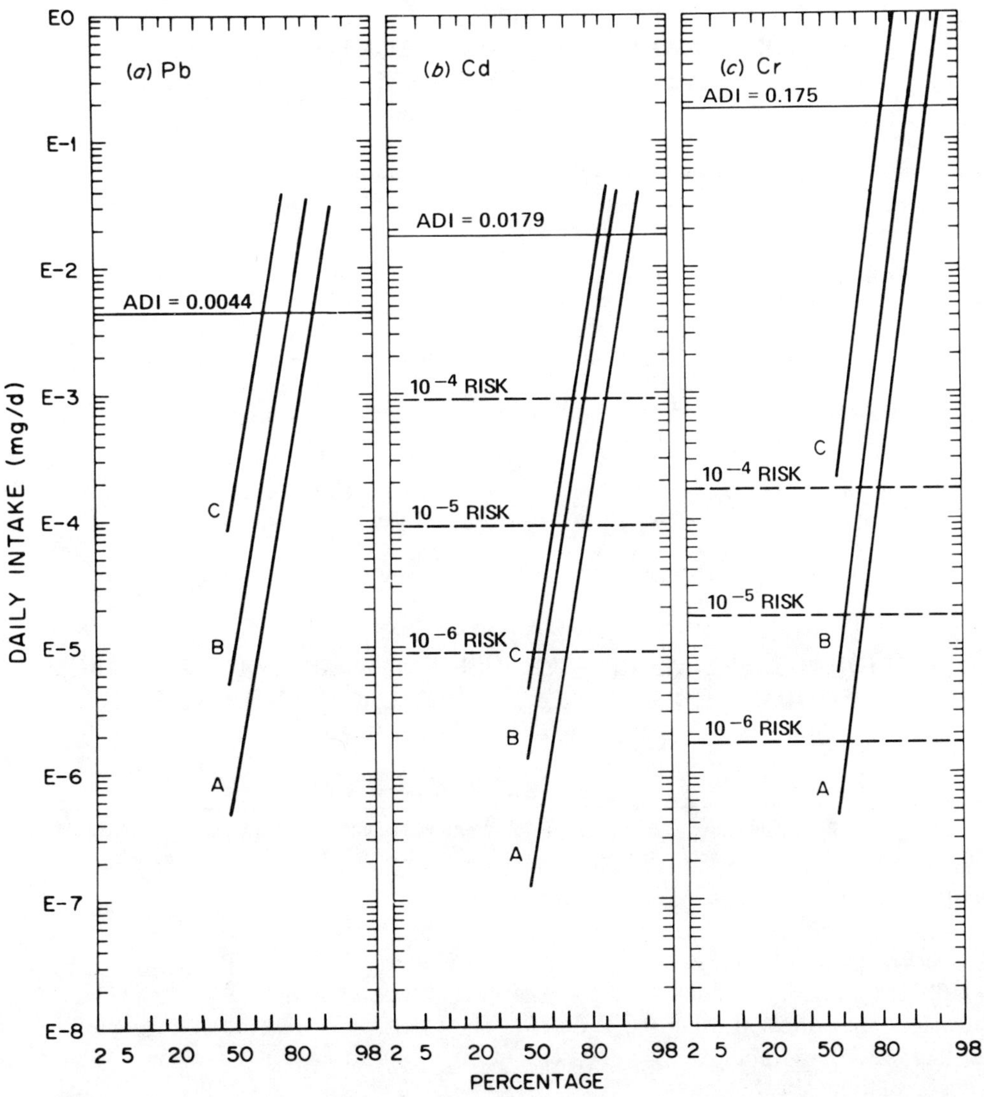

FIGURE 4. Daily intake (milligram per day) distributions of lead (a), cadmium (b), and chromium (c) for individuals in the vicinity of A, a 10 × 10^6 Btu/h incinerator, average individual exposure; B, a 150 × 10^6 Btu/h incinerator, average individual exposure; and C, a 150 × 10^6 Btu/h incinerator, maximum individual exposure.

risk is 2.7E − 10. The average risks associated with maximum measured emissions for Cd and nickel are 3.6E − 8 and 3.5E − 8, respectively. For Cr, the excess cancer risk associated with the maximum measured emissions rate is 1.0E − 6.

The mean of the distribution of the highest individual exposures is 3.4 times that of the average individual exposure. For the mean highest individual exposure values, the associated maximum cancer risks range from 91.E − 10 for beryllium to 3.5E − 6 for Cr.

2. *Noncarcinogenic Risk*

Table 10 presents the average daily intake ranges of 12 toxic metals measured by Trenholm et al.[12] as fractions of their respective ADI values. Even assuming the mean highest sector values for individual exposures, no values equal or exceed the ADI. Under these conditions, the highest Pb intake is estimated at 2% of the ADI, while intakes of other metals are all lower.

Table 8
RANGE OF INDIVIDUAL EXPOSURES TO HAZARDOUS METALS FROM HAZARDOUS WASTE INCINERATION AT FIVE SITES

	Average individual exposure	Maximum individual exposures
Antimony	<3.9E − 6 to 3.9E − 4	<1.3E − 5 to 1.3E − 3
Arsenic	6.2E − 6	2.0E − 5
Barium	1.0E − 6 to 3.9E − 5	3.3E − 6 to 1.3E − 4
Beryllium	<3.1E − 8 to 6.4E − 7	<1.1E − 7 to 2.2E − 6
Cadmium	1.1E − 6 to 2.8E − 5	3.6E − 6 to 9.6E − 5
Chromium	2.1E − 6 to 1.5E − 4	7.3E − 6 to 5.2E − 4
Lead	<5.3E − 6 to 2.3E − 3	<1.8E − 5 to 7.9E − 3
Mercury	1.4E − 6	4.8E − 6
Nickel	<1.3E − 6 to 1.9E − 4	4.4E − 6 to 6.3E − 4
Selenium	<1.5E − 5 to 1.5E − 3	5.1E − 5 to 5.0E − 3
Silver	3.3E − 7 to 2.1E − 6	1.1E − 6 to 7.3E − 6
Thallium	<1.3E − 6 to 3.9E − 6	<4.4E − 6 to 1.3E − 5

From Travis, C. C. and Munro, N. B., unpublished paper, 1986.

Table 9
INDIVIDUAL LIFETIME CANCER RISK FROM ARSENIC, BERYLLIUM, CADMIUM, CHROMIUM, AND NICKEL EMISSIONS AT FIVE INCINERATORS

		Cancer risk per person	
	Emission ranges (g/sec)	Average individual exposure	Maximum individual exposure
Arsenic	3.0E − 4	1.4E − 8	4.8E − 8
Beryllium	<1.6E − 6 to 3.3E − 5	1.3E − 11 to 2.7E − 10	4.5E − 11 to 9.3E − 10
Cadmium	5.5E − 5 to 1.5E − 3	1.4E − 9 to 3.6E − 8	4.6E − 9 to 1.2E − 7
Chromium	1.1E − 4 to 7.8E − 3	1.4E − 8 to 1.0E − 6	4.9E − 8 to 3.5E − 6
Nickel	<6.7E − 5 to 9.5E − 3	<2.5E − 10 to 3.5E − 8	8.3E − 10 to 1.2E − 7

From Travis, C. C. and Munro, N. B., unpublished paper, 1986.

V. SUMMARY

The use of probability distributions for obtaining quantitative estimates of variability in metal emissions and associated health risks changes radically the interpretation one would otherwise make of the likely hazards associated with metals from hazardous waste incineration. The distributions indicate that a significant number of persons may be exposed to high levels of risk, particularly for the largest incinerators and maximally exposed individuals.

Consider the distributions for cancer risk associated with Cd and Cr emissions. The rightmost curves in Figure 4 (line A) show the cancer risk distributions for persons receiving the average daily exposure to Cd and Cr in the vicinity of a medium-sized (10×10^6 Btu/h) incinerator. Under these assumptions, for an incinerator chosen at random in the U.S. and burning average waste containing Cd or Cr, there is a 29% chance that persons around it would encounter a lifetime cancer risk of 10^{-6} or more from Cd and a 32% chance of such levels of risk from Cr. Under the worst case assumptions, for a randomly chosen incinerator there is a 48% chance that persons around it would encounter a 10^{-6} or greater risk of cancer from Cd and a 58% chance of such levels of risk from Cr.

Table 10
AVERAGE DAILY INTAKE[a] OF HEAVY METALS RELEASED IN TESTS OF FIVE HAZARDOUS WASTE INCINERATORS (PRESENTED AS A FRACTION OF ADI)

Metal	ADI (mg/d)	Intake as a fraction of ADI
Antimony	0.292	1.5E − 7 to 1.5E − 5
Arsenic	0.07	9.5E − 7
Barium	0.179	6.2E − 7 to 2.5E − 6
Beryllium	0.00071	5.2E − 7 to 1.0E − 5
Cadmium	0.0179	6.9E − 7 to 1.1E − 4
Chromium (VI)	0.175	1.4E − 7 to 1.0E − 5
Lead	0.0045	1.3E − 5 to 5.9E − 3
Mercury	0.02	8.0E − 7
Nickel	1.46	1.0E − 8 to 1.5E − 6
Selenium	0.70	2.5E − 7 to 2.4E − 5
Silver	0.016	2.4E − 7 to 1.5E − 6
Thallium	0.0373	4.0E − 7 to 1.2E − 6

[a] Based on a mean individual exposure of 1.95E − 2 $\mu g/m^3$ per g/sec unit emission rate derived from PEER Report.[8]

From Travis, C.C. and Munro, N. B., unpublished paper, 1986.

The mean values for average daily intakes of Pb, Cd, and Cr lie several orders of magnitude below the ADI values. Even with very large variability as indicated by large values for σ_{ab}, if a medium size (10×10^6 Btu/h) incinerator burning the same waste year long is chosen at random, there is a 5 to 9 $^1/_2$% chance that persons around it will receive a daily intake greater than the ADI for a given metal. However, when the case of a randomly selected 150×10^{10} Btu/h incinerator is examined, the chance of persons in its vicinity receiving the ADI level or more rises to 13% for Cr, 17% for Pb, and 12% for Cd. Looking at the worst case scenario of a large incinerator and maximally exposed individual, the chance of persons in the vicinity receiving the ADI level or more reaches 22% for Cr, 28% for Pb, and 16% for Cd.

The calculations assume no background exposure to these metals from other sources. If the background from other sources is taken into account, the potential health impacts are of much more concern, particularly for Pb.

In past assessments[10,12] where no measure of variability was available, it appeared that metal emissions from incineration of hazardous wastes were not likely to pose health problems even under worst case conditions. However, the use of techniques to obtain quantitative estimates of variability show that excessive exposures to several metals can occur for significant fractions of people living in the vicinity of hazardous waste incinerators, especially those of the largest size. It appears that the use of highly efficient emission control systems is desirable, particularly so for large incinerators burning wastes containing higher levels of heavy metals.

REFERENCES

1. **Aitchison, J. and Brown, J. A. C.,** *The Lognormal Distribution,* Cambridge University Press, New York, 1969.
2. **Bowers, J. F., Bjorklund, J. R., and Cheney, C. S.,** Industrial Source Complex (ISC) Dispersion Model User's Guide, Vol. 1, EPA-450/4-79-030, U.S. Environmental Protection Agency, Research Triangle Park, NC, 1979.
3. **Haus, S. and Gerstein, R.,** Hazardous Waste Stream Metal Concentrations and Emissions, working paper prepared by the MITRE Corporation for the Office of Solid Waste, U.S. Environmental Protection Agency, Washington, D.C., November 1983.
4. **Kelly, K. E.,** Health Risk Assessment of Hazardous Waste Incinerator Stack Emissions, Ph.D. dissertation, Columbia University School of Public Health, New York, 1985.
5. **Kelly, K. E.,** Health risk assessment of hazardous waste incinerator stack emissions, in *Hazardous Waste and Hazardous Chem.,* 3(4), 367, 1986.
6. **Midwest Research Institute, Inc.,** Inhalation Exposure Methodology (IEM) Modeling of Hazardous Waste Incinerators, prepared for the U.S. Environmental Protection Agency, Washington, D.C., 1983.
7. **O'Donnell, F. R. and Holton, G. A.,** An automated methodology (IEM) for assessing inhalation exposure to hazardous waste incineration emissions, in Incineration and Treatment of Hazardous Waste Proc. 9th Annu. Res. Symp. Fort Mitchell, Kentucky, May 24, 1983, U.S. Environmental Protection Agency, Cincinnati, OH, 1983.
8. PEER Consultants, Inc., Supporting Documentation for the RCRA Incinerator Regulations, 40 CFR #264, Subpart O — Incinerators, U.S. Environmental Protection Agency, Washington, D.C., 1984.
9. **Speer, D. R. and Waite, D. A.,** Statistical Distributions as Applied to Environmental Surveillance Data, in Proc. 3rd Environmental Protection Conf., Report CONF-750967, U.S. Energy Research and Development Administration, NTIS, Springfield, VA, 1975, 117.
10. **Travis, C. C., Etnier, E. L., Holton, G. A., O'Donnell, F. R., Hetrick, D. M., Dixon, E., and Harrington, E. S.,** Inhalation Pathway Risk Assessment of Hazardous Waste Incineration Facilities, ORNL/TM-9096, Oak Ridge National Laboratory, Oak Ridge, TN, 1984.
11. **Travis, C. C. and Munro, N. B.,** Risk Assessment of Heavy Metal Emissions from Hazardous Waste Incineration Facilities, unpublished paper, 1986.
12. **Trenholm, A., Gorman, P., and Jungclaus, G.,** Performance Evaluation of Full-Scale Hazardous Waste Incinerators, report prepared by Midwest Research Institute (MRI) for the U.S. Environmental Protection Agency, Cincinnati, OH, 1984.
13. Health Effects Assessment Summary for 300 Hazardous Organic Constitutents in Support of Regulatory Impact Analysis of the Land Disposal Branch (LDB) and Interim Final Incinerator Regulations of the Technical Branch (TB) of the Office of Solid Wastes, OSWER/EPA, Environmental Criteria Assessment Office, U.S. Environmental Protection Agency, Cincinnati, OH, 1982.
14. Summary of Current Acceptable Daily Intakes for Various Pollutants, Environmental Criteria Assessment Office, U.S. Environmental Protection Agency, Cincinnati, OH, 1983.
15. Health Assessment Document for Inorganic Arsenic, Final Report, EPA-600/8-83-021F, U.S. Environmental Protection Agency, Research Triangle Park, NC, 1984.
16. **Vogel, G., Holberger R., and Haus, S.,** Composition of Hazardous Waste Streams Currently Incinerated — Working Paper, Office of Solid Waste, U.S. Environmental Protection Agency, Washington, D.C., 1983.

Chapter 7

PRODUCTS OF INCOMPLETE COMBUSTION

I. INTRODUCTION

The prime objective of hazardous waste incineration is to thermally destroy organic compounds. Ideally, hazardous organic compounds entering an incinerator's firing chamber will degrade into simpler forms such as carbon dioxide and water.[15] Unfortunately, the efficiency of the combustion process is not 100%; consequently, some waste compounds pass through the incineration process completely unchanged, others are partially broken down to simpler compounds, and, in some cases, new compounds form outside the heating zone due to the recombination of molecular fragments. We have previously discussed the health effects associated with emission of waste compounds that pass through the incineration process completely unchanged. However, possible health effects associated with compounds formed during the incineration process have not been addressed.

Analysis of incinerator stack gas samples gives evidence of the emission of organics not originally detected in the waste stream. These organics have been termed products of incomplete combustion (PICs). Due to studies which report that PIC emissions from well-operated hazardous waste incinerators are very low[5,18] stack gas emissions of PICs are currently unregulated, although the Environmental Protection Agency (EPA) has proposed regulations[20] under the Resource Conservation and Recovery Act which limit PIC emissions to 0.01% of principal organic hazardous constituents (POHC) input rates.[15] It is the purpose of this chapter to survey limited existing data on PIC emissions and reach a preliminary evaluation of their potential human health impact.

II. PRODUCTS OF INCOMPLETE COMBUSTION (PICs)

PICs have been defined as the EPA Appendix VIII compounds present in stack gas, but not found above 100 μg/g in the waste.[18] From a health risk perspective, the proper definition of PCIs should be *all* compounds present in stack gas which are not present in the waste stream. If the Appendix VIII list is used as a defining element for PICs, then hazardous compounds absent from this EPA list cannot be classified as PICs. They also cannot be classified as POHCs, since POHCs, by definition, must also be Appendix VIII compounds. The result is that a large percentage of stack emission remains unclassified.[15] The total quantity of organic compounds excluded from the categories of POHC and PIC is possibly larger than the combined constituents that are listed and analyzed. Moreover, these compounds may be more toxic and pose higher environmental risks than the Appendix VIII compounds.

A. Sources

Organic compounds classified as PICs have several possible origins:

1. The detected PICs may be new organic compounds formed during the incineration process due to recombination of molecular fragments outside the combustion zone
2. The detected PICs may have been present in the waste at concentrations slightly below the 100 μg/g cut-off level used in analyzing the waste feed, and/or
3. The PICs detected in stack gas may have been brought into the incineration system from noncombustion sources such as off-gases from scrubber water used in control devices, ambient air introduced into the system, or auxiliary fuel feed

Table 1
PICs FOUND IN TRIAL BURN STACK GAS

PIC	Number of sites	Stack exhaust concentrations (μg/m³)
Benzene	6	12—670
Chloroform	5	1—1330
Bromodichloromethane	4	3—32
Dibromochloromethane	4	1—12
Bromoform	3	0.2—24
Naphthalene	3	5—100
Chlorobenzene	3	1—10
Tetrachloroethylene	3	0.1—2.5
1,1,1-Trichloroethane	3	0.1—1.5
Hexachlorobenzene	2	0.5—7
Methylene chloride	2	2—27
o-Nitrophenol	2	25—50
Phenol	2	4—22
Toluene	2	2—75
Bromochloromethane	1	14
Carbon disulfide	1	32
Methylene bromide	1	18
2,4,6-Trichlorophenol	1	110
Bromomethane	1	1
Chloromethane	1	3
Pyrene	1	1
Flouranthene	1	1
Dichlorobenzene	1	2—4
Trichlorobenzene	1	7
Methyl ethyl ketone	1	3
Diethyl phthalate	1	7
o-chlorophenol	1	2—22
Pentachlorophenol	1	6
2,4 dimethyl phenol	1	1—21

From Trenholm, A., Gorman, P., and Jungclaus, G., Vol. 1 to 4, EPA-600/2-84-181a-3, Industrial Environmental Research Laboratory, U.S. Environmental Protection Agency, Cincinnati, OH, November 1984.

From a risk assessment perspective, it is immaterial whether PICs detected in stack gas are truely new compounds formed during the incineration process or are compounds originally present in the waste stream, scrubber water, or auxiliary fuel. Regardless, the compounds are released to the atmosphere as a result of the incinerator process, and their human health impact has not been accounted for in the POHC risk analysis.

Table 1 lists PICs found in stack effluents at the eight EPA trial burns. The hypothesis that most PICs are the result of a recombination of molecular fragments is supported by the appearance of many of the same organic compounds in almost all trial burn stack gas samples.[18] Many of the compounds on this list are suspected human carcinogens.

III. DIOXINS AND FURANS

Public and federal agency concern about PICs intensified in the late 1970s when emissions of highly toxic chlorinated dioxins and furans (two families of PICs) were discovered at both municipal and hazardous waste incineration facilities. The term polychlorinated dibenzo-*p*-dioxins (dioxins for short) refers to a group of organic compounds composed of two benzene rings which are connected by a pair of oxygen atoms and surrounded by chlorine

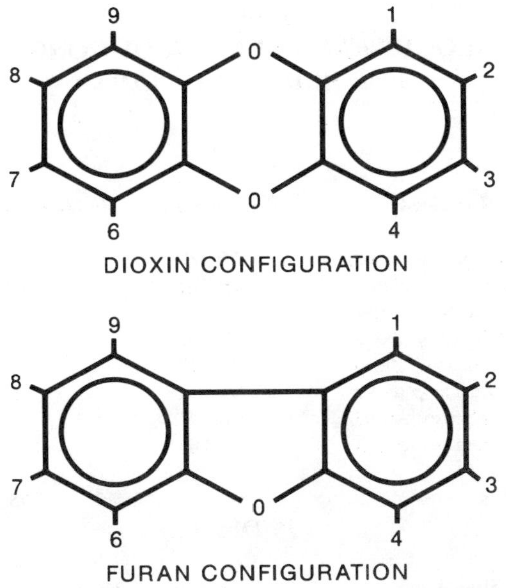

FIGURE 1. Dioxin and furan configurations.

atoms in eight possible substituent positions (Figure 1). This chlorination can potentially create 75 different congeners. Congeners are compounds that have the same molecular formula, but different patterns of substitution or molecular geometry. The varying physicial and chemical properties of each congener arises from the number and position of its chlorine atoms.[22] Polychlorinated dibenzo-*p*-dioxins are referred to by the acronym PCDDs.

The 75 PCDD congeners are divided into eight homologues according to the number of chlorine atoms they contain (from one to eight). Every member of the homologus group tetrachlorodibenzo-*p*-dioxin (TCDD), for example, contains four chlorine atoms. A member of this homologue, 2,3,7,8-tetrachlorodibenzo-*p*-dioxin (2,3,7,8-TCDD) is the focus of special attention. The 2,3,7,8 nomenclature designates the positions where the four chlorine atoms connect to the two benzene rings (see Figure 1); based on previous studies, the 2,3,7,8-TCDD congener is considered the most toxic of the chlorinated dioxins and is the isomer correlated most frequently with human exposure and potential health risks. This congener is often used interchangeably in the literature with the general term "dioxin".

Furans comprise another family of organic PIC compounds, the members of which possess structural similarities to dioxins and are chemically referred to as dibenzofurans, chlorinated dibenzofurans, or polychlorinated dibenzofurans. The two benzene rings of a furan nucleus are connected by one oxygen atom and surrounded by chlorine atoms in eight possible substituent positions (Figure 1). This chlorination can potentially create 135 furan congeners with different physical and chemical properties. All 135 furans are referred to by the acronym PCDFs. Table 2 lists the dioxin and furan homologues and their respective number of congeners.

A. Sources

Of the 210 dioxin and furan congeners, none are manufactured or have any known use.[22] Despite this fact, their presence is widespread in our environment. Besides trace contamination of air, water, and soil,[8] PCDDs and PCDFs have been detected in human adipose tissue samples collected from the general population.[14] This widespread presence in the environment is the result of their inadvertent formation during certain manufacturing processes and as combustion byproducts. Specific sources of dioxins and furans in the envi-

Table 2
HOMOLOGUES AND NUMBER OF CONGENERS FOR PCDDs AND PCDFs

PCDDs

Homologue	Acronym	Number of congeners
Monochlorodibenzo-*p*-dioxin	MCDD	2
Dichlorodibenzo-*p*-dioxin	DCDD	10
Trichlorodibenzo-*p*-dioxin	TrCDD	14
Tetrachlorodibenzo-*p*-dioxin	TCDD	22
Pentachlorodibenzo-*p*-dioxin	PnCDD	14
Hexachlorodibenzo-*p*-dioxin	HxCDD	10
Heptachlorodibenzo-*p*-dioxin	HpCDD	2
Octachlorodibenzo-*p*-dioxin	OCDD	1
Total PCDDs		75

PCDFs

Homologue	Acronym	Number of congeners
Monochlorodibenzofuran	MCDF	4
Dichlorodibenzofuran	DCDF	16
Trichlorodibenzofuran	TrCDF	28
Tetrachlorodibenzofuran	TCDF	38
Pentachlorodibenzofuran	PnCDF	28
Hexachlorodibenzofuran	HxCDF	16
Heptachlorodibenzofuran	HpCDF	4
Octachlorodibenzofuran	OCDF	1
Total PCDFs		135

From Barnes, D. G., presented at the Symp. Energy from Biomass and Wastes VII, Institute of Gas Technology, Lake Buena Vista, FL, January 24 to 28, 1983.

ronment include (1) manufacturing processes (production of chlorophenols and their derivatives), (2) municipal waste incineration, (3) other combustion processes (i.e., hazardous waste incineration), (4) chemical disposal sites, and (5) photochemical processes.

Traces of PCDDs and PCDFs can form when chlorinated hydrocarbons are combusted.[22] They have been detected in particulate emissions from both municipal and industrial incinerators. A recent study[10] at a hazardous waste incinerator in Sweden has correlated stack releases of dioxins and furans with total chlorine input to the incinerator.

Based on previous studies, dioxin emissions from combustion sources appear to be affected by the following factors:[23]

1. PCDD in waste feed — The possibility of finding dioxins in the feed of a combustion process is enhanced by the widespread uses of commercial products such as pesticides and wood preservatives in which PCDDs can be found as trace contaminants.
2. Dioxin precursors in the feed — Three classes of chemical compounds have been isolated and studied as potential precursors of PCDD emissions from combustion: chlorinated phenols, chlorinated benzenes, and PCBs.
3. Waste feed chlorine content — Chlorine is a necessary element in the formation of PCDD, and increased chlorine concentrations in the feed can enhance the occurrence of PCDD formation.[10]

4. Combustion temperature — Temperatures between 500 and 800°C are assumed to promote PCDD formation, while destruction of PCDDs is expected at temperatures above 800°C. Combustion temperature is a function of fuel Btu content, available O_2, and fuel processing. Thus, the conditions of municipal waste incinerator combustion (low temperatures, high moisture, low Btu fuel, lack of supplemental fuel, and poor waste mixing) are conducive to and considered a significant source of PCDD formation. Hazardous waste incinerators and high efficiency boilers, on the other hand, have combustion design features that are less conducive to the formation and emission of dioxins.
5. Residence time — In general, PICs are less likely to occur in incinerators with long residence times at high temperatures. The residence times necessary for decomposition of PCDDs shorten as temperature increases.
6. Oxygen availability — Poor combustion conditions (leading to the formation of PCDDs) result from low ratios of air to fuel and/or poor mixing of air and fuel.
7. Processing of waste feed — Waste feeds which are processed into small, homogeneous particles enhance the mixing of air and fuel and require shorter residence times. Wastes composed of larger particles may result in poorly mixed oxygen and fuel and require longer residence times. Waste mixes with high moisture content can also decrease combustion efficiency. A lack of processing detracts from complete combustion and, consequently, increases the likelihood of PCDD formation.
8. Use of supplemental fuel — Combustion sources firing supplemental fuel with high Btu content are assumed to be unlikely producers of PCDDs. Combustion sources burning low Btu wastes that do not utilize supplemental fuel are high potential sources of PCDDs.

B. Health Effects

Knowledge of the environmental activity of dioxins is limited. PCDDs are highly toxic; the LD_{50} level of 2,3,7,8-TCDD for guinea pigs, for example, is 0.6 μg/kg. Other PCDD congeners are regarded as less toxic, although the LD_{50} values remain in the microgram per kilogram range.[22] Exposure to 2,3,7,8-TCDD can occur via inhalation, dermal penetration, or gastrointestinal absorption. Although inhalation exposure to 2,3,7,8-TCDD could result from inhalation of mist, dust, or other contaminated particulate matter, this type of exposure is unlikely due to the low vapor pressure of this congener.[22]

Symptoms of humans exposed to 2,3,7,8-TCDD are similar to effects observed in exposed animals. Exposure in both cases may lead to altered liver function, lipid metabolism, porphyria cutanea tarda, neurotoxicity, and pathologic changes in hematologic parameters. Exposure to 2,3,7,8-TCDD may also produce chloracne and hyperpigmentation in humans. It is important to note that although some toxicity symptoms are attributed to PCDDs, other toxic effects may derive from the other chemicals in which 2,3,7,8-TCDD exists as a contaminant.[22]

Sources seem to agree that human exposures to 2,3,7,8-TCDD lead to the symptoms listed above, whether the dose is acute or chronic. One estimate[22] of the cumulative minimal toxic dose for man is 0.1 μg/kg. Epidemiological studies, based largely on populations exposed to Herbicide Orange in Vietnam, demonstrate both short-term, subsiding, and long-term, persistent toxic effects following exposure to 2,3,7,8-TCDD.

Evidence suggesting that 2,3,7,8-TCDD is a mutagen is limited and contradictory.[22] Carcinogenicity evidence for this compound is supported by chronic animal bioassays. These studies indicate that 2,3,7,8-TCDD is a potent animal carcinogen; the most potent analyzed by the EPA to date (see Table 3, Chapter 4). Limited epidemiologic studies indicate that this PCDD congener may also be a human carcinogen. However, human exposures to 2,3,7,8-TCDD are in conjunction with other chemicals present in herbicides, making the isolated evaluation of the dioxin's carcinogenic effects in humans a speculative one.

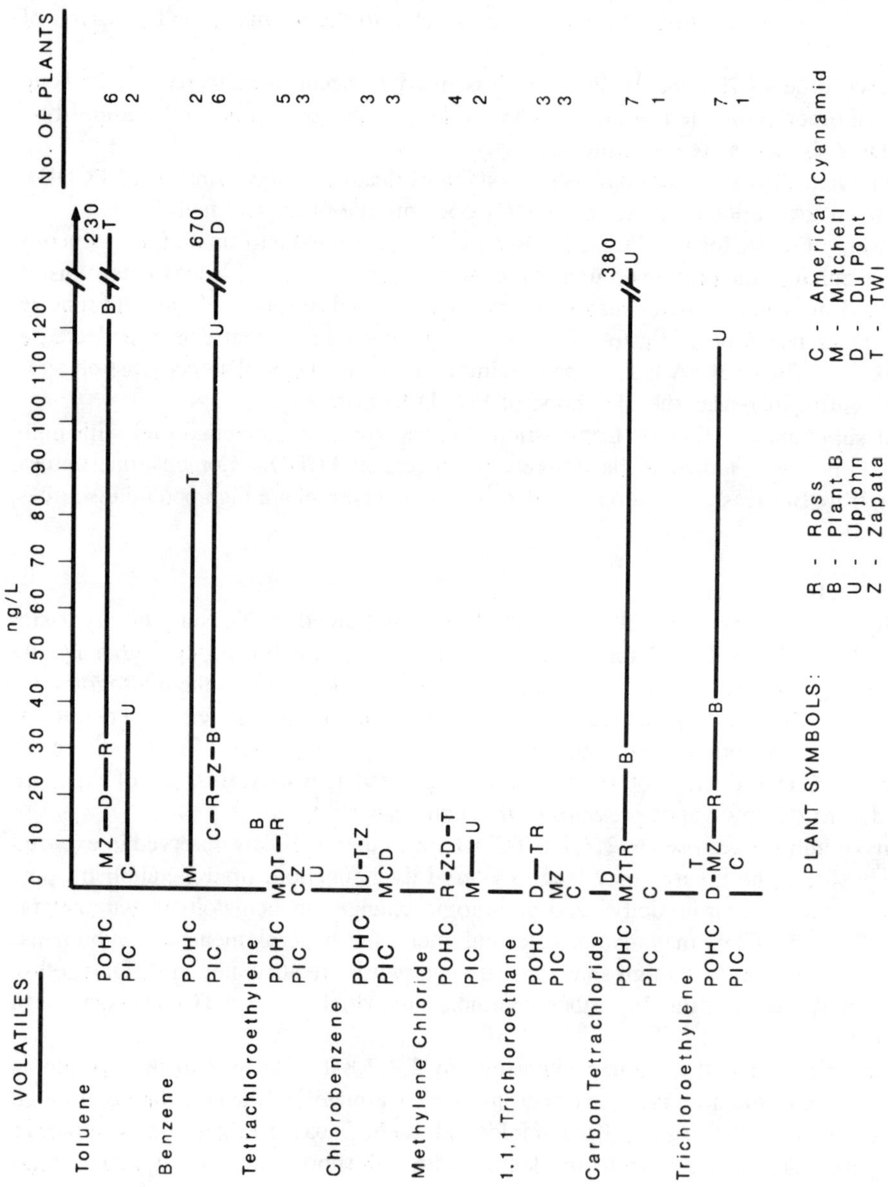
VOLATILES
ng/L
No. OF PLANTS
Toluene
Benzene
Tetrachloroethylene
Chlorobenzene
Methylene Chloride
1,1,1 Trichloroethane
Carbon Tetrachloride
Trichloroethylene
POHC
PIC
PLANT SYMBOLS:
R - Ross
B - Plant B
U - Upjohn
Z - Zapata
C - American Cyanamid
M - Mitchell
D - Du Pont
T - TWI
A

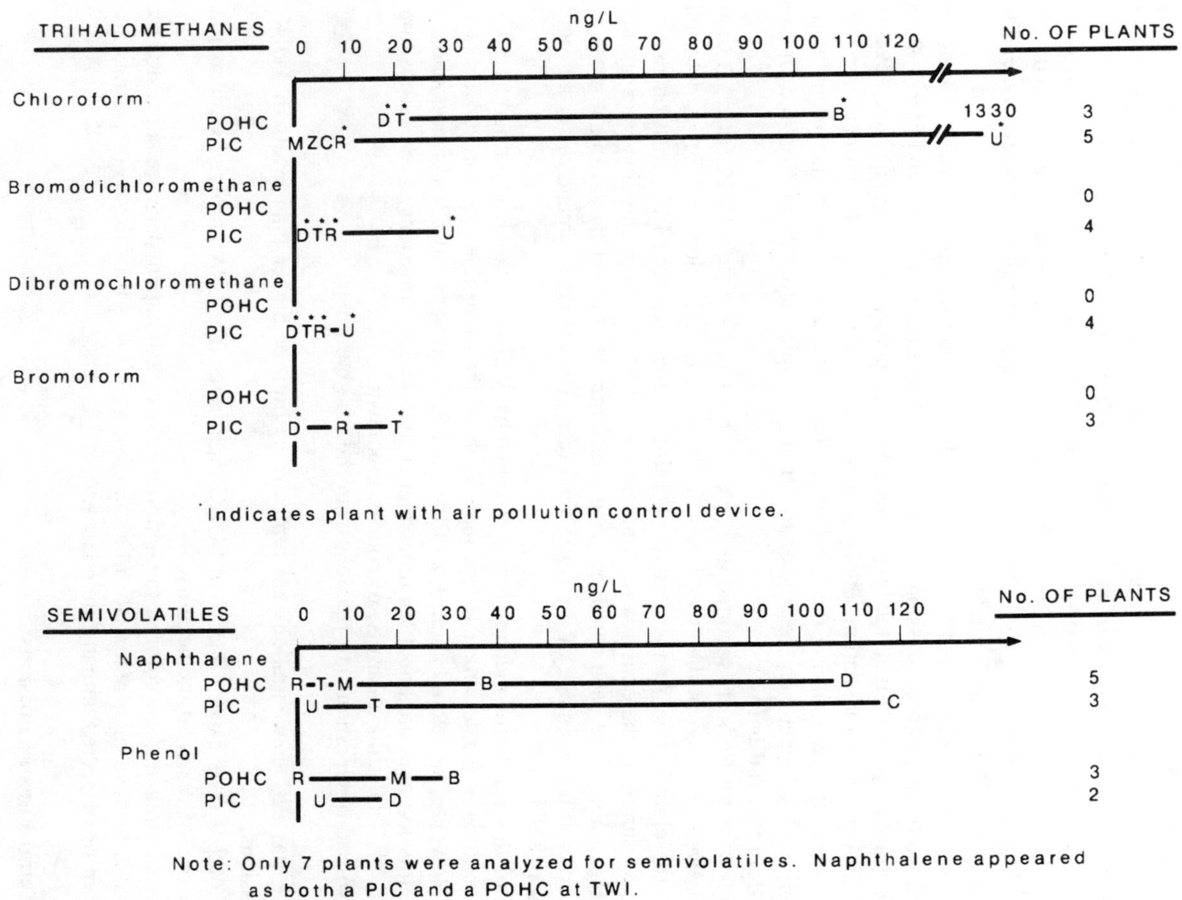

B

FIGURE 2. PIC and POHC stack gas concentrations at eight trial burn facilities.

IV. TRIAL BURN STACK EMISSIONS

The following sections summarize stack gas emissions of PICs and dioxins and furans from several trial burns. PIC emissions data are primarily from Trenholm et al.,[18] PCDD and PCDF emissions data are from EPA[21,24] and MAIN.[9]

A. Products of Incomplete Combustion (PIC) Emissions

The current data base for incinerator emissions of PICs is very limited.[15] The most complete data are from EPA-sponsored trial burns at eight incinerators.[18] Different organic waste streams (non-PCB) were burned in each of the eight trial burns. PICs were defined as any Appendix VIII compound emitted in the stack gas which was not present in the waste in concentrations greater than 100 μg/g. Because the stack gas samples were analyzed only for Appendix VIII compounds, PIC emission rates obtained from the trial burn data tend to underestimate total PIC emission rates.

Trenholm et al.[18] estimate that the measured PICs account for approximately 1% of the total unburned hydrocarbons emissions. Figure 2 illustrates compared PIC and POHC stack gas concentrations at the eight facilities. Because the incinerator operators had advance notice of EPA personnel visitation, the incinerators are likely to have been operating at or near maximum efficiency during the test firings. Thus, PIC emissions in hazardous waste incinerator stack effluent may be higher under more typical conditions than the figures reported from the trial burns.[15]

Emission rates for PICs are reported in Table 3 for each of the eight trial burns.[24] A total of 35 PICs were detected in the stack gases, with benzene and chloroform being detected in five of the eight burns. At the bottom of Table 3, the measured PIC emission rate is calculated by summing the emission rates for individual PICs. These emission rates are not directly comparable since each trial burn used a different waste feed rate. Table 4 presents PIC emissions as a percentage of the waste feed rate. The average percentage for all eight trial burns is 8.9E − 4.

Results of the trial burns indicate that the Appendix VIII compounds identified as PICs are being emitted at a rate which is less than $^1/_{10}$th of the allowable rate for POHCs (see Table 5). However, since only about 1% of the total PICs are Appendix VIII compounds,[24] total PIC emission rates are about ten times higher than POHC emissions. The composition of the total PIC emission stream is, of course, unknown.

Several broad observations are possible from the trial burn data:

1. Stack gas PIC concentrations are typically as high or higher than levels of POHC emissions.
2. PIC output rates exceed 0.01% of POHC input rates (the EPA proposed standard) at only one of the eight incinerators.
3. Benzene, toluene, chloroform, tetrachloroethylene, and naphthalene are suggested by the data to have high potential to be manifest as PICs. All of these compounds are known or suspected human carcinogens.

B. Dioxin and Furan Emissions

The current data base on emissions of dioxins and furans from hazardous waste incinerators is limited.[15] As with PIC emissions, the most complete quantitative data are from the EPA-sponsored trial burns at eight incinerators.[18] Six of the eight trial burn incinerators burned non-PCB waste and two burned wastes containing PCBs.

1. Dioxin and Furan Emissions from Combustion of Non-PCB Waste

The data base for stack concentrations of dioxins and furans from hazardous waste incinerators tested in the EPA trial burns is summarized in Table 6. This table indicates the

Table 3
PIC EMISSIONS RATES (μg/min) AT LAND-BASED INCINERATORS FOR NON-PCB ORGANIC WASTES

PIC	Trial burn 1	2	3	4
1,1,1-Trichloroethane				1.6E + 1
1,2,4-Trichlorobenzene		1.5E + 3		
2,4,6-Trichlorophenol			5.9E + 3	
2,4-Dimethyl phenol				
2,6-Toluene diisocyanate			2.2E + 2	
Benzene	9.8E + 3		3.6E + 3	5.6E + 2
Bromoform	8.2E + 3			
Carbon disulfide	2.7E + 4			
Carbon tetrachloride				
Chlorobenzene				
Chloroform	7.3E + 3		8.3E + 4	5.3E + 1
Dibromochloromethane	4.0E + 3		1.1E + 3	
Dichlorobromomethane	4.5E + 3		2.2E + 3	
Diethyl phthalate			3.8E + 2	
Dimethyl phthalate		1.7E + 2		
Flouranthene	1.9E + 3			
Hexachlorobenzene		1.5E + 3	2.3E + 1	
Methyl bromide	1.2E + 3			
Methyl chloride	4.1E + 3			
Methyl ethyl ketone			1.7E + 2	
Methylene bromide	1.5E + 4			
Methylene bromochloride	1.6E + 4			
Methylene chloride			8.0E + 2	
Naphthalene			2.7E + 2	
Pentachlorophenol			3.5E + 2	
Phenol			2.3E + 2	
Pyrene	8.3E + 2			
Tetrachloroethylene			1.5E + 2	1.6E + 1
Toluene			3.3E + 3	
Trichloroethylene				
o-Dichlorobenzene		8.4E + 2		
o-Chlorophenol			4.8E + 2	
o-Nitrophenol			1.3E + 3	
o-Dichlorobenzene		1.1E + 3		
o-Dichlorobenzene		6.0E + 2		
Total emission rates	9.8E + 4	5.7E + 3	1.0E + 5	6.6E + 2
Average waste feed rates (kg/min)	98.77	41.14	1.96	8.74
1,1,1-Trichloroethane	2.5E + 1	8.2E + 0		
1,2,4-Trichlorobenzene				
2,4,6-Trichlorophenol				
2,4-Dimethyl phenol		6.3E + 2		
2,6-Toluene diisocyanate				
Benzene	5.0E + 2		2.5E + 5	
Bromoform			7.6E + 1	1.5E + 3
Carbon disulfide				
Carbon tetrachloride	6.8E + 1			
Chlorobenzene	8.0E + 1	8.0E + 1	1.9E + 3	
Chloroform	3.2E + 2	1.0E + 2		
Dibromochloromethane			3.6E + 2	4.3E + 2

Table 3 (continued)
PIC EMISSIONS RATES (μg/min) AT LAND-BASED INCINERATORS FOR NON-PCB ORGANIC WASTES

	Trial burn			
PIC	1	2	3	4
Dichlorobromomethane			1.1E + 3	5.8E + 2
Diethyl phthalate				
Dimethyl phthalate				
Flouranthene				
Hexachlorobenzene				
Methyl bromide				
Methyl chloride				
Methyl ethyl ketone				
Methylene bromide				
Methylene bromochloride				7.4E + 2
Methylene chloride		1.5E + 2		
Naphthalene	6.9E + 3			1.9E + 3
Pentachlorophenol				
Phenol			8.1E + 3	
Pyrene				
Tetrachloroethylene	1.2E + 1			
Toluene	2.4E + 2			
Trichloroethylene	4.3E + 2			
o-Dichlorobenzene				
o-Chlorophenol				
o-Nitrophenol	2.5E + 3			
o-Dichlorobenzene				
o-Dichlorobenzene				
Total emission rate	1.1E + 4	9.7E + 2	2.6E + 5	7.1E + 3
Average waste feed rate (kg/min)	2.52	9.64	22.57	7.26

From U.S. Environmental Protection Agency, Washington, D.C., March 1985 and Trenholm, A., Gorman, P., and Jungclaus, G., Vol. 1 to 4, EPA-600/2-84-181a-3, Industrial Environmental Research Laboratory, U.S. Environmental Protection Agency, Cincinnati, OH, November 1984.

dioxin/furan detection limits used where applicable and the likely dioxin precursors in the waste feed and/or fuel.[23] These data are of limited use in quantifying dioxins and furan emissions at hazardous waste incinerators since emission rates are not given.

Trenholm et al.[18] measured concentrations of dioxins and furans in stack gas at six of the eight trial burn incinerators burning non-PCB waste. Dioxins are detected at one incinerator, with a maximum concentration of 0.02 ng/l. Furans are measured at three sites, with a maximum concentration of 0.06 ng/l. Table 7 presents the stack gas concentrations for dioxins and furans at two of the non-PCB waste trial burns.[18] In order to obtain dioxin and furan emission rates, it is necessary to multiply these data by the volumetric velocity of the stack gas. A representative number for volumetric velocity is 1.2E + 3 l/sec.[24] Using this rough estimate, emission rates for dioxin and furans are also presented in Table 7.

Oberg et al.[10] have also reported dioxin and furan emission rates at a Swedish hazardous waste incinerator burning non-PCB waste. The emission rates are presented in Table 8 and are somewhat larger than those measured at the EPA trial burns.

Table 4
PIC EMISSION RATES (% OF WASTE FEED) AT LAND-BASED INCINERATORS FOR NON-PCB ORGANIC WASTE

PIC	Average emission rate (%)	Trial burn			
		1	2	3	4
1,1,1,-Trichloroethane	4.1E − 7				2.7E − 6
1,2,4-Trichlorobenzene	4.6E − 7		3.7E − 5		
2,4,6-Trichlorophenol	3.7E − 5			3.0E − 4	
2,4-Dimethyl phenol	8.2E − 7				
2,6-Toluene diisocyanante	1.3E − 6			1.0E − 5	
Benzene	1.7E − 4	9.1E − 6		1.8E − 4	7.6E − 5
Bromoform	7.0E − 6	8.3E − 6			
Carbon disulfide	3.4E − 6	2.7E − 5			
Carbon tetrachloride	3.4E − 7				
Chlorobenzene	1.6E − 6				
Chloroform	5.3E − 4	7.4E − 6		4.2E − 3	8.0E − 6
Dibromochloromethane	8.6E − 6	4.8E − 6		5.7E − 5	
Dichlorobromomethane	1.7E − 5	4.6E − 6		1.2E − 4	
Diethyl phthalate	2.4E − 6			1.9E − 5	
Dimethyl phthalate	5.1E − 8		4.1E − 8		
Flouranthene	1.4E − 7	1.1E − 6			
Hexachlorobenzene	5.9E − 7		3.5E − 6	1.2E − 6	
Methyl bromide	1.6E − 7	1.3E − 6			
Methyl chloride	5.1E − 7	4.1E − 6			
Methyl ethyl ketone	1.1E − 8			8.8E − 6	
Methylene bromide	2.0E − 6	1.6E − 5			
Methylene bromochloride	3.3E − 6	1.5E − 5			
Methylene chloride	5.3E − 6			4.1E − 5	
Naphthalene	3.9E − 5			1.4E − 5	
Pentachlorophenol	2.2E − 6			1.7E − 5	
Phenol	6.0E − 6			1.2E − 5	
Pyrene	1.1E − 7	6.4E − 7			
Tetrachloroethylene	1.3E − 6			7.8E − 6	2.3E − 6
Toluene	2.1E − 5			1.5E − 4	
Trichloroethylene	2.1E − 6				
o-Dichlorobenzene	2.6E − 7		2.5E − 6		
o-Chlorophenol	3.1E − 6			2.4E − 5	
o-Nitrophenol	2.1E − 5			6.8E − 5	
o-Dichlorobenzene	3.3E − 7		2.6E − 6		
o-Dichlorobenzene	1.8E − 7		1.5E − 6		
Total	8.9E − 4	10.0E − 5	1.4E − 5	5.3E − 3	5.8E − 5
1,1,1-Trichloroethane	4.1E − 7	1.8E − 6	8.5E − 8		
1,2,4-Trichlorobenzene	4.6E − 7				
2,4,6-Trichlorophenol	3.7E − 5				
2,4-Dimethyl phenol	8.2E − 7		6.5E − 6		
2,6-Toluene diisocyanate	1.3E − 6				
Benzene	1.7E − 4	2.0E − 5		1.1E − 3	
Bromoform	7.0E − 6			3.4E − 7	4.8E − 5
Carbon disulfide	3.4E − 6				
Carbon tetrachloride	3.4E − 7	2.8E − 6			
Chlorobenzene	1.6E − 6	3.2E − 6	8.3E − 7	8.5E − 6	
Chloroform	5.3E − 4	1.3E − 5	1.0E − 6		
Dibromochloromethane	8.6E − 6			1.6E − 6	5.9E − 6
Dichlorobromomethane	1.7E − 5			5.1E − 6	7.9E − 6
Diethyl phthalate	2.4E − 6				

Table 4 (continued)
PIC EMISSION RATES (% OF WASTE FEED) AT LAND-BASED INCINERATORS FOR NON-PCB ORGANIC WASTE

PIC	Average emission rate (%)	Trial burn 1	Trial burn 2	Trial burn 3	Trial burn 4
Dimethyl phthalate	5.1E − 8				
Flouranthene	1.4E − 7				
Hexachlorobenzene	5.9E − 7				
Methyl bromide	1.6E − 7				
Methyl chloride	5.1E − 7				
Methyl ethyl ketone	1.1E − 8				
Methylene bromide	2.0E − 6				
Methylene bromochloride	3.3E − 6				1.1E − 5
Methylene chloride	5.3E − 6		1.6E − 6		
Naphthalene	3.9E − 5	2.7E − 4			2.6E − 5
Pentachlorophenol	2.2E − 6				
Phenol	6.0E − 6			3.7E − 5	
Pyrene	1.1E − 7				
Tetrachloroethylene	1.3E − 6	4.8E − 7			
Toluene	2.1E − 5	9.4E − 6			
Trichloroethylene	2.1E − 6	1.7E − 5			
o-Dichlorobenzene	2.6E − 7				
o-Chlorophenol	3.1E − 6				
o-Nitrophenol	2.1E − 5	9.9E − 5			
o-Dichlorobenzene	3.3E − 7				
o-Dichlorobenzene	1.8E − 7				
Total	8.9E − 4	4.4E − 4	1.0E − 5	1.2E − 3	9.8E − 5

From U.S. Environmental Protection Agency, Washington, D.C., March 1985 and Trenholm, A., Gorman, P., and Jungclaus, G., Vol. 1 to 4, EPA-600/2-84-181a-3, Industrial Environmental Research Laboratory, U.S. Environmental Protection Agency, Cincinnati, OH, November 1984.

Table 5
CALCULATED VALUES FOR PIC EMISSIONS AS A PERCENT OF POHC INPUT

Plant	PIC output (g/min)	POHC input (g/min)	%
Ross	0.0900	9800	0.00092
Plant B (semivolatiles only)	0.0046	100	0.0044
Upjohn	0.092	780	0.012
Zapata	0.00062	27	0.0023
American Cyanamid	0.011	1200	0.00091
Mitchell	0.00090	310	0.00029
DuPont	0.26	8300	0.0031
TWI	0.0076	1200	0.00061

Note: $\% = \frac{\text{PIC output (g/min)}}{\text{POHC input (g/min)}} \times 100$. High blank values for volatile PICs.

From Trenholm, A., Gorman, P., and Jungclaus, G., Vol. 1 to 4, EPA-600/2-84-181a-3, Industrial Environmental Research Laboratory, U.S. Environmental Protection Agency, Cincinnati, OH, November 1984.

Table 6
DIOXIN/FURAN EMISSIONS MEASURED AT HAZARDOUS WASTE INCINERATORS

Facility	Precursors	Detection	Isomer	Isomer limits
Rollins, TX	Waste chemicals	0.8 ng	TCDF	4.82 ng/Nm^3
ENSCO, AR	Waste chemicals	0.2 ng	TCDF	0.56 ng/Nm^3
Rollins, TX	Waste chemicals	0.4 ng	TCDD	2.48 ng/Nm^3
ENSCO, AR	Waste chemicals	0.07 ng	TCDD	0.72 ng/Nm^3
New Jersey	Silvex herbicide	NA	2,3,7,8-TCDD	Not detected
New Jersey	Silvex herbicide	NA	PCDD	Not detected
Mitchell, NC	Organic (0.4—1% Cl)	NA	TCDF	0.7 ng/m^3
DuPont, LA	Liq/Sol (0.05—16% Cl)	NA	TCDF	Not detected
Upjohn, TX	Liq/Gas (21% Cl)	NA	TCDD	Not detected
Upjohn, TX	Liq/Gas (21% Cl)	NA	TCDF	6.8 ng/m^3
Ross, OH	Liq Org (0.02—6.2% Cl)	NA	TCDD	Not detected
American Cyanamid, WV	Aniline waste (0.15% Cl)	NA	TCDD	Not detected
Ross, OH	Liq Org(0.02—6.2% Cl)	NA	TCDD	Not detected
American Cyanamid, WV	Aniline Waste (0.15% Cl)	NA	TCDF	Not detected
American Cyanamid, WV	Aniline waste (0.15% Cl)	NA	TCDD	Not detected
American Cyanamid, WV	Aniline waste (0.15% Cl)	NA	TCDF	Not detected
Upjohn, TX	Liq/gas (21% Cl)	NA	TCDD	Not detected
Mitchell, NC	Liq org (0.4—1% Cl)	NA	TCDD	Not detected
DuPont, LA	Liq/sol (0.05—16% Cl)	NA	TCDD	Not detected
Mitchell, NC	Liq org (0.4—1% Cl)	NA	TCDF	Detected
Ross, OH	Liq org (0.02—6.2% Cl)	NA	TCDF	Not detected
Mitchell, NC	Liq org (0.4—1% Cl)	NA	TCDD	Not detected
DuPont, LA	Liq/sol (0.05—16% Cl)	NA	TCDF	Not detected
Ross, OH	Liq org (0.02—6.2% Cl)	NA	TCDF	Not detected
Upjohn, TX	Liq/gas (21% Cl)	NA	TCDF	Detected
DuPont, LA	Liq/sol (0.05—16% Cl)	NA	TCDD	Not detected

Note: NA = not applicable; Liq = liquid; Sol = solid; and Org = organic.

Adapted from U.S. Environmental Protection Agency, EPA-450/4-84-014b, Research Triangle Park, N.C., 1984.

Table 7
STACK SAMPLES OF DIOXINS AND FURANS AT TWO NON-PCB TRIAL BURN FACILITIES[a]

	Plant B stack		Mitchell stack	
	Conc (ng/l)	Emission rate (ng/sec)	Conc (ng/l)	Emission rate (ng/sec)
TCDF	0.047	57.81	0.0009	1.11
PCDF	0.023	28.29	0.0012	1.48
TCDD	0.021	25.83	<0.0003	<0.37
PCDD	<0.002	<2.46	<0.0003	<0.37

[a] Concentrations from Reference 18.

Table 10
DIOXIN AND FURAN EMISSIONS (% OF WASTE FEED) AT LAND-BASED TRIAL BURN INCINERATORS BURNING PCB WASTE[9,15,20]

Dioxin/ furan	Average emission rate (%)	SCA trial burns					Rollins 1980	ENSCO 1980
		10	11	12	13	14		
Di CDF	1.2E − 7	2.1E − 7	1.6E − 7	0.0E + 0	9.4E − 8	1.3E − 7		
Hepta CDF	1.3E − 9	0.0E + 0	0.0E + 0	0.0E + 0	1.5E − 11	6.4E − 9		
Hexa CDF	2.5E − 8	0.0E + 0	0.0E + 0	0.0E + 0	0.0E + 0	1.2E − 7		
Mono CDF	4.4E − 8	0.0E + 0	1.9E − 7	0.0E + 0	0.0E + 0	3.2E − 8		
Octa CDD	1.3E − 7	3.2E − 7	0.0E + 0	1.2E − 7	1.5E − 7	3.3E − 8		
Octa CDF	6.4E − 8	0.0E + 0	1.3E − 8	0.0E + 0	3.0E − 7	1.2E − 7		
Pent CDF	4.0E − 8	2.1E − 8	5.3E − 8	3.9E − 8	3.1E − 8	5.7E − 8		
TCDD	3.4E − 8	2.7E − 8	1.1E − 7	5.0E − 8	3.1E − 8	1.9E − 8	1.5E − 9	2.8E − 10
TCDF	1.1E − 8	1.1E − 8	1.3E − 8	0.0E + 0	6.1E − 9	1.9E − 8	2.4E − 8	3.5E − 9
Tri CDF	7.5E − 8	1.8E − 7	0.0E + 0	5.0E − 8	1.1E − 8	1.3E − 7		
Total (%)	5.4E − 7	7.8E − 7	5.4E − 7	2.6E − 7	6.2E − 7	5.5E − 7	2.5E − 8	3.8E − 9

Table 8
EMISSION RATES (ng/sec) FOR DIOXINS AND FURANS AT A HAZARDOUS WASTE INCINERATOR[10]

Isomer	Conc
PCDD	112
Tetra-CDD	16
Penta-CDD	64
PCDF	400
Tetra-CDF	64
Penta-CDF	144

Table 9
DIOXIN AND FURAN EMISSION RATES (μg/sec) AT LAND-BASED TRIAL BURN INCINERATORS BURNING PCB WASTE[9,15,20]

	SCA trial burns					Rollins	ENSCO
Dioxin/furan	10	11	12	13	14	1980	1980
Octa CDD	4.0E − 1	0.0E + 0	1.0E − 1	5.2E − 1	2.7E − 1		
Di CDF	2.7E − 1	3.2E − 1	0.0E + 0	3.2E − 1	1.1E + 0		
tri CDF	2.3E − 1	0.0E + 0	4.1E − 2	3.6E − 2	1.1E + 0		
Octa CDF	0.0E + 0	2.6E − 2	0.0E + 0	1.0E + 0	1.0E − 1		
TCDD	3.4E − 2	2.2E − 1	4.1E − 2	1.0E − 1	1.6E − 1	1.0E − 8	1.7E − 9
Mono CDF	0.0E + 0	1.7E − 1	0.0E + 0	0.0E + 0	2.6E − 1		
Pent CDF	2.7E − 2	1.0E − 1	3.1E − 1	1.0E − 1	4.7E − 1		
TCDF	1.4E − 2	2.6E − 2	0.0E + 0	2.1E − 2	1.6E − 1	1.6E − 7	2.1E − 8
Hexa CDF	0.0E + 0	0.0E + 0	0.0E + 0	0.0E + 0	1.0E + 0		
Hepta CDF	0.0E + 0	0.0E + 0	0.0E + 0	5.2E − 5	5.3E − 2		
Total	9.8E − 1	1.1E + 0	2.1E − 1	2.1E + 0	4.6E + 0	1.7E − 7	2.3E − 8

Note: CDD = chlorinated dibenzo-*p*-dioxins, and CDF = chlorinated dibenzofurans.

2. *Dioxin and Furan Emissions from Combustion of PCB Waste*

Table 9 compiles dioxin and furan emission rates from the three trial burn facilities burning PCB waste in Arkansas, Texas, and Illinois. At the bottom of Table 9, the total dioxin and furan emission rate is calculated by summing the emission rates for individual dioxins and furans. These emission rates are not directly comparable since each trial burn used a different waste feed rate.

Table 10 lists dioxin and furan emissions as a percentage of waste feed at the same three facilities. The average total dioxin and furan emission rate over all the trial burn data, as a percentage of the waste feed rate, is 5.4E − 7. This value may not be representative since it is based on data from one incinerator (the SCA Chemical Services facility in Chicago, IL); total PCDD and PCDF emissions were not measured at the Rollins and Energy Systems Company (ENSCO) trial burn facilities.

V. PIC RISK ASSESSMENT

In order to estimate the worst case exposures that could result if all the unidentified hydrocarbons emitted in incinerator stacks are Appendix VIII hazardous compounds, we consider three liquid injection incinerators burning pesticide-related waste (with the same

Table 11
EXCESS CANCER RISK FROM PICs IDENTIFIED IN INCINERATOR STACK EFFLUENT[17,18]

PIC	Excess cancer risk $(mg/kg/d)^{-1}$
Benzene	0.052
Chloroform	0.18
Tetrachloroethylene	0.04
2,4,6-Trichlorophenol	0.02
Hexachlorobenzene	1.68

scenario as the three sites, S-1, S-2, and S-3, described in Chapter 5) for which only 1% of the emissions is identified as PICs and POHCs, and the other 99% is unidentified PICs. The following sections describe a PIC health risk assessment based on these assumptions.

A. Stack Emissions

The data on the magnitude of (identified) PIC output as a percentage of POHC input (see Table 5) will be used as the starting point for estimating the emission rate for unidentified PICs.[17] The average PIC output, according to Trenholm et al.[18] is 0.0031% of POHC input. Multiplying this percentage by the pesticide-related waste throughput for a 150×10^6 Btu/h incinerator yields 8.6×10^5 g/year of observed PIC emissions. If the PIC output measured by MRI (Midwest Research Institute) represents 1% of total PIC output (based on a worst case assumption), an upper limit to total PIC releases from the incinerator is estimated at 8.6×10^7 g/year. This compares with 2.8×10^6 g/year in total POHC stack emissions (assuming 99.99% destruction and removal efficiency [DRE] for POHCs).

B. Population Exposure

The methodology for estimating atmospheric transport of PICs utilizes the automated inhalation exposure methodology (IEM).[11] This air dispersion model is a slightly modified version of the Industrial Source Complex Long-Term Model (ISCLTM) developed for the EPA.[2] Population distribution for a 100 km radius around three incinerators is similar to that assumed in the risk assessment for POHCs described in Chapter 5. The assessment areas are divided into sector segments consisting of 20 concentric circles about the origin and 16 radial direction vectors. The total exposure for a sector segment is calculated by multiplying the annual-average ground-level air concentration at the segment centroid by the number of persons located in the segment. At site S-1, for example, the entire population assumed is 0.45×10^6. The estimated PIC emissions of 8.6×10^7 g/year from a 150×10^6 Btu/h liquid injection incinerator result in a collective population exposure of 3.4×10^3 person — micrograms per meter^3 at that site.

C. Health Risk for Products of Incomplete Combustion (PICs)

To estimate the potential health risks for PICs, those compounds with cancer potency data are listed in Table 11. Their toxicities fall in the range of 0.02 to 1.7 $(mg/kg/d)^{-1}$. Assuming the entire PIC emission stream has a toxicity of 1.7 $(mg/kg/d)^{-1}$ (probably a worst case assumption), an upper bound estimate of the excess cancers per year in the entire population at site S-1 (0.45×10^6 persons) from PICs released by an LI-150 incinerator is 0.02. Excess lifetime cancer risks from PIC emissions to the maximally exposed individual

Table 12
WORST CASE ESTIMATE OF EXCESS LIFETIME CANCER RISK TO THE MAXIMALLY EXPOSED INDIVIDUAL FROM PICs AT LIQUID INJECTION INCINERATORS BURNING PESTICIDE-RELATED WASTE

Site	LI-1	LI-10	LI-150
S-1	1.3E − 6	1.3E − 5	1.9E − 4
S-2	2.2E − 6	2.2E − 5	3.3E − 4
S-3	3.3E − 6	3.3E − 5	5.0E − 4

From Travis, C. C., Etnier, E. L., Holton, G. A., O'Donnell, F. R., Hetrick, D. M., Dixon, E., and Harrington, E. S., ORNL/TM-9096, Oak Ridge National Laboratory, Oak Ridge, TN, October 1984.

Table 13
SUMMARY OF DIOXIN AND FURAN CANCER RISK ESTIMATES FOR TRIAL BURN SITES[9,15,20]

	Arkansas	Texas	Chicago
Population risk	1.1E − 1	6.6E − 1	3.3E + 0
Maximum individual risk	8.2E − 6	1.2E − 5	—
Expected individual risk	1.2E − 6	2.6E − 7	1.5E − 6

for three sites burning pesticide-related wastes are given in Table 12. Given that we made worst case assumptions in deriving the estimates in Table 12, the human health risk from PIC emissions is probably small.

It is important to note, however, that a complete characterization of PIC emissions from incinerators is not available. It can be argued that most of the unidentified hydrocarbon emissions are compounds of low molecular weight (e.g., alkanes), which either have no significant health effects derived through inhalation or which have low impact on human health compared to the effect of the identified PICs. The counter argument is that PIC emissions may contain quantities of highly toxic compounds. Until the complete PIC emission stream is characterized, we will not know the answer with certainty.

VI. DIOXIN AND FURAN RISK ASSESSMENT

Dioxin and furan risk assessments have been performed for trial burns at three operating U.S. facilities burning PCBs: ENSCO in El Dorado, AR; Rollins Environmental Services in Deer Park, TX; and SCA Chemical Services in Chicago, IL.

A. Health Risk Estimates

Dioxin and furan emission data from Table 9 for ENSCO, Rollins, and SCA Chemical Services can be used to predict maximum ground level pollutant concentrations. These predictions are of the same order of magnitude, 10^{-14} to 10^{-15} g/m^3, for all three facilities.[21]

The major assumptions used to calculate carcinogenic risk are as follows: exposure is for 70 years; human exposure is through inhalation; exposed individuals inhale 20 m^3 of air per day, retain 65% of the inhaled dioxins and furans, and weigh 70 kg. The EPA estimates the unit risk factor associated with lifetime inhalation exposure to air containing 1 ng/m^3 of 2,3,7,8-TCDD as 0.091. (All dioxins and furans are considered to be as toxic as 2,3,7,8-TCDD for our assessment.)

Table 13 summarizes the dioxin/furan inhalation risk for the three facilities. The largest population risk occurs at the Chicago incinerator, with results indicating about three cancer

Table 14
RELATIVE TOXICITIES OF PCDDs AND PCDFs

PCDD/PCDF	Swiss[a]	Grant, Olie, and Commoner[b]	New York State[c]	Ontario[d]	FDA[e]	California[f]	EPA 1981[g]	EPA present[h]
Mono thru di	0	0	0	0	0	0	0	0
tri	0	0	0	1	0	0	0	0
2378-TCDD	1	1	1	1	1	1	1	1
other TCDDs	0.01	0	0	0.01	0	0	1	0.01
2378-PeCDD	0.1	0.1	1	1	0	1	0	0.5
other PeCDDs	0.1	0	0	0.01	0	0	0	0.005
2378-HxCDD	0.1	0.1	0.03	1	0.02	1	0	0.04
other HxCDDs	0.1	0	0	0.01	0.02	0	0	0.0004
2378-HpCDD	0.01	0.01	0	1	0.005	1	0	0.001
other HpCDDs	0.01	0	0	0.01	0.005	0	0	0.00001
OCDD	0	0	0	0	<0.00001	1	0	0
2378-TCDF	0.1	0.1	0.33	0.02	0	1	0	0.1
other TCDFs	0.1	0	0	0.0002	0	0	0	0.001
2378-PeCDF	0.1	0.1	0.33	0.02	0	1	0	0.1
other PeCDFs	0.1	0	0	0.0002	0	0	0	0.001
2378-HxCDF	0.1	0.1	0.01	0.02	0	1	0	0.01
other HxCDFs	0.1	0	0	0.0002	0	0	0	0.0001
2378-HpCDF	0.1	0.01	0	0.02	0	1	0	0.001
other HpCDFs	0	0	0	0.0002	0	0	0	0.00001
OCDF	0	0	0	0	0	0	0	0

[a] From Reference 16.
[b] From References 3, 6, and 12.
[c] From Reference 4.
[d] From Reference 13.
[e] From Reference 19.
[f] From Reference 7.
[g] From Reference 20.
[h] EPA currently recommends.

deaths over a 70-year period. The risk to the maximally exposed individual for the Arkansas facility is estimated as 8.2×10^{-6} (less than one excess cancer incidence per 122,000 persons). At the Texas facility, the risk is 1.2×10^{-5} (less than one excess cancer incidence per 83,000 exposed persons). The risk to the maximally exposed individual was not estimated for the Chicago facility.

B. 2,3,7,8-Tetrachlorodibenzo-*p*-Dioxin (2,3,7,8-TCDD) Toxicity Equivalence Factors

Emissions of dioxins and furans contain a mixture of the 75 chlorodibenzo-*P*-dioxin (CDD) and 135 chlorodibenzo-*P*-furan (CDF) congeners. Historically, it has been assumed that all these congeners were as carcinogenic as 2,3,7,8-TCDD. An alternative approach is the use of short-term biological tests as indirect indicators of the toxicity of a mixture of CDDs and CDFs. The type of short-term tests used in this approach include acute toxicity, carcinogenicity, reproductive effects, receptor binding, enzyme induction, and *in vitro* cell transformation. Based on these tests, the EPA and other organizations have estimated the relative toxicity of the major congeners of CDD and CDF (see Table 14). For ease of comparison, the data are normalized to unity for 2,3,7,8-TCDD. For example, the EPA estimates that 2,3,7,8 PeCDD is one half as toxic as 2,3,7,8-TCDD.

A risk assessment of dioxin and furan emissions, using the toxicity equivalence factor (TEF) approach, involves the following steps:[25]

1. Determination of the concentration of the CDDs and CDFs in the emission stream
2. Multiplication of the congener concentrations in the sample by the TEFs in Table 14 to express the concentration in terms of 2,3,7,8-TCDD equivalents
3. Summation of the products in Step 2 to obtain the total 2,3,7,8-TCDD equivalents in the emissions
4. Determination of the human exposure to the mixture, expressed in terms of 2,3,7,8-TCDD equivalents
5. Multiplication of the 2,3,7,8-TCDD equivalents exposure estimates obtained in Step 4 by the carcinogenic potency estimates for 2,3,7,8-TCDD to obtain an estimate of risk associated with the mixture

VII. SUMMARY

When hazardous organic wastes are incinerated, new organic compounds can form outside the heating zone due to the recombination of molecular fragments. These organics have been termed PICs. The current data base for incinerator emissions of PICs is very limited.[15] The most complete data are from the EPA-sponsored trial burns at eight incinerators. Results of the trial burns indicate that the Appendix VIII compounds identified as PICs are being emitted at a rate which is less than $^{1}/_{10}$th of the allowable rate for POHCs (see Table 5). However, since only about 1% of the total PICs are Appendix VIII compounds,[24] total PIC emission rates are about ten times higher than POHC emissions. The composition of the total PIC emission stream is, of course, unknown. To estimate possible health effects of PIC emissions, we assume the entire PIC emission stream (both Appendix VIII and non-Appendix VIII compounds) has an average toxicity of 1.7 $(mg/kg/day)^{-1}$. Under these assumptions, the average individual lifetime cancer risk is 2.7×10^{-6}. Thus, it does not appear that PICs, considered as a whole, will pose an unreasonable risk to surrounding populations. However, to verify this conclusion, more research is needed to determine PIC emission stream composition and emission rates.

Despite the fact that the above calculations show that total PIC emissions may not pose a health problem, these estimates were based on an assumed average toxicity of PICs. In reality, some small fraction of the PIC emission stream may possess extremely high toxicity, an example is dioxin and furan emissions from PCB incineration. Our risk calculations for PCB incineration show lifetime risk to the maximally exposed individual is about 1.2×10^{-5}.

To date dioxin and furan emissions have only been monitored at PCB incinerators equipped with air pollution control devices (APCDs). A recent study[10] at a hazardous waste incinerator in Sweden has correlated stack releases of dioxins and furans with total chlorine input to the incinerator. Thus, dioxin/furan emissions can occur at incinerators which are not burning PCBs. Dioxin/furan emission rates should be determined for incinerators burning non-PCB organic wastes, particularly those without APCDs.

REFERENCES

1. **Barnes, D. G.,** Dioxin production from combustion of biomass and waste, presented at the Symp. Energy from Biomass and Wastes VII, Institute of Gas Technology, Lake Buena Vista, FL, January 24 to 28, 1983.
2. **Bowers, J. F., Bjorklund, J. R., and Cheney, C. S.,** Industrial Source Complex (ISC) Dispersion Model User's Guide, Vol. 1, EPA-450/4-79-030, U.S. Environmental Protection Agency, Research Triangle Park, NC, 1979.
3. **Commoner, B., et al.,** Environmental and economic analysis of alternative municipal solid waste disposal technologies. I. An assessment of the risks due to emissions of chlorinated dioxins and dibenzofurans from proposed New York City incinerators, May 1, 1984.
4. **Eadon, G., et al.,** Comparisons of chemical and biological data on soot samples from the Binghamton State Office Building, unpublished report, 1982.
5. **Gorman, P. G. and Ananth, K. P.,** Trial Burn Protocol Verification at a Hazardous Waste Incinerator, Industrial Environmental Research Laboratory, U.S. Environmental Protection Agency, Cincinnati, OH, 1982.
6. **Grant, D. L.,** Proc. 12th Annu. Workshop Pesticide Residues Analysis, Winnipeg, Canada, 1977, 251.
7. **Gravitz, N.,** Interim Guidelines for Acceptable Exposure Levels in Office Settings Contaminated with PCB and PCB Combustion Products, Epidemiological Studies Section, California Department of Health Services, Sacramento, CA, November 1, 1983.
8. **Hutzinger, O., Blumich, M. T., Berg, M., and Olie, K.,** Sources and fate of PCDDs and PCDFs: an overview, *Chemosphere,* 14(6/7), 581, 1985.
9. Chas. T. Main, Inc., Health Risk Assessment of the Incineration of Polychlorinated Biphenyls, prepared for SCA Chemical Services, Inc. by Chas. T. Main, Boston, MA, 1983.
10. **Oberg, T., Aittola, J. P., and Bergstrom, J. G. T.,** Chlorinated aromatics from the combustion of hazardous waste, *Chemosphere,* 14(2), 215, 1985.
11. **O'Donnell, F. R. and Holton, G. A.,** An automated methodology (IEM) for assessing inhalation exposure to hazardous waste incineration emissions, Incineration and Treatment of Hazardous Waste, Proceedings of the 9th Annual Research Symposium at Fort Mitchell, KY, May 2 to 4, 1983, U.S. Environmental Protection Agency, Cincinnati, OH, 1984.
12. **Olie, K., Berg, M., and Hutzinger, O.,** Formation and fate of PCDD and PCDF from combustion processes, *Chemosphere,* 12, 627, 1983.
13. Ontario Government, Chlorinated dioxins and chlorinated dibenzofurans. Ambient air guideline, Health Studies Service, Ministry of Labor, Ontario, Canada, December 16, 1982.
14. **Ryan, J. J., Lizotte, R., and Lau, B. P. Y.,** Chlorinated dibenzo-*p*-dioxins and chlorinated dibenzofurans in Canadian human adipose tissue, *Chemosphere,* 14(6/7), 697, 1985.
15. **Science Advisory Board,** Report on the Incineration of Liquid Hazardous Wastes, Environmental Effects, Transport and Fate Committee, U.S. Environmental Protection Agency, Washington, D.C., April 1985.
16. Swiss Government (Bundesamt fur Umweltschutz, Bern), Environmental Pollution Due to Dioxins and Furans from Chemical Rubbish Incineration Plants, Schriftenreighe Umweltschutz, No. 5, 1982.
17. **Travis, C. C., Etnier, E. L., Holton, G. A., O'Donnell, F. R., Hetrick, D. M., Dixon, E., and Harrington, E. S.,** Inhalation Pathway Risk Assessment of Hazardous Waste Incineration Facilities, ORNL/TM-9096, Oak Ridge National Laboratory, Oak Ridge, TN, October 1984.
18. **Trenholm, A., Gorman, P., and Jungclaus, G.,** Performing Evaluation of Full-Scale Hazardous Waste Incinerators, Vols. I-V, EPA-600/2-84-181a-3, Industrial Environmental Research Laboratory, U.S. Environmental Protection Agency, Cincinnati, OH, November 1984.

19. U.S. Department of Health and Human Services, Levels of Concern for Hexa-(HCDD), Hepta-(HpCDD) and Octachlorodibenzo-*p*-Dioxins (OCDD) in Chickens and Eggs, memorandum, U.S. Department of Health and Human Services, April 29, 1983.
20. Background Document, Identification and Listing of Hazardous Waste Under RCRA, Subtitle C, Section 3001, U.S. Environmental Protection Agency, Washington, D.C., January 1981.
21. Interim evaluation of health risks associated with emission of tetrachlorinated dioxins from municipal waste resource recovery facilities, Office of the Administrator, U.S. Environmental Protection Agency, Washington, D.C., November 19, 1981.
22. Health Assessment Document for Polychlorinated Dibenzo-*p*-Dioxins, EPA-600/8-84-014A, U.S. Environmental Protection Agency, Washington, D.C., 1984.
23. National Dioxin Study Tier 4 — Combustion Sources. Initial Review and Testing Options, EPA-450/4-84-014b, U.S. Environmental Protection Agency, Research Triangle Park, NC, 1984.
24. Assessment of Incineration as a Treatment Method for Liquid Organic Hazardous Wastes, U.S. Environmental Protection Agency, Washington, D.C., March 1985.
25. Interim Procedures for Estimating Risks Associated with Exposures to Mixtures of Chlorinated Dibenzo-*p*-Dioxins and Dibenzofurans (CDDs and CDFs), U.S. Environmental Protection Agency, Washington, D.C., 1986.

Chapter 8

INCINERATION OF POLYCHLORINATED BIPHENYLS

I. INTRODUCTION

Polychlorinated biphenyls (PCBs) are derivations of the compound biphenyl in which the hydrogen atoms have been replaced with one to ten chlorine atoms. PCBs are stable compounds, resistant to oxidation, reduction, and other chemical reactions. Due to this stability, plus excellent dialectric and heat transfer properties, PCBs have been used internationally in many industrial and consumer applications.[21] Unfortunately, PCBs are now classified among the most persistent and widespread environmental pollutants[13] with documented pathobiological responses upon exposure.[1]

In 1972, a Federal interagency task force concluded that PCBs were highly persistent, could be found in all parts of the environment, could bioaccumulate to relatively high levels in fish, and could have serious adverse effects on human health. As D'ltri and Kamrin[6] state: "The story of PCBs is one of major global contamination by a group of highly useful chemicals which now appear in the tissues of most living creatures as well as in the air, water, soils, and sediments over most of the earth."

Because Congress recognized PCBs as a threat to exposed humans, PCB regulations were included in the 1976 Toxic Substance Control Act (TSCA). This Act, designed to limit human exposure to all toxics, gives the Environmental Protection Agency (EPA) the mandate to implement a program controlling PCBs. The EPA decreed that "the manufacture, processing, and distribution in commerce of PCBs at concentrations of 50 ppm or greater, and PCB items with PCB concentrations of 50 ppm or greater, present an unreasonable risk of injury to health within the U.S."[5] Thus, commercial PCB activity was phased out beginning May 31, 1979 (effective July 2, 1979). "While . . . prohibited in the U.S., the size of the PCBs environmental burden in waters, sediments, soils, disposal sites, and in deployed transformers and other containers of PCBs is very large, having been estimated by NAS[11] to be 82×10^6 kg."[6] The final "PCB Ban Rule" specifies approved landfills and incineration as the only acceptable methods for disposal of PCBs and PCB-contaminated materials (50 to 500 ppm) and incineration only for PCB substances 500 ppm and above. The present chapter surveys the available literature concerning the health risks of incineration of PCBs.

II. POLYCHLORINATED BIPHENYLS (PCBs)

PCBs are a group of related chemical compounds with the general formula $C_{12}H_xCl_{10}$-$_x$ (see Figure 1). These substances consist of phenyl rings with 1 to 10 chlorine and hydrogen atoms attached in substituent positions, forming a possible 210 compounds and isomers.[4]

A. Sources

PCBs were first manufactured in the U.S. in 1929. They have been used extensively by the electric utility industry in power distribution systems. Because PCBs are chemically stable and resistant to fire, as well as having excellent dialectric properties, they have been employed in capacitators and transformers. PCBs have also been utilized in a variety of other industrial products including: heat transfer systems, hydraulic systems, gas turbines, vacuum pumps, fire retardants, plasticizers, and carbonless copy paper.[7] These uses were widespread until the PCB ban came into effect in 1979.

From 1929 to 1975, approximately 6.4×10^{11} g of PCBs were manufactured in the U.S. by the Monsanto Industrial Chemicals Corporation.[7] This production accounts for approx-

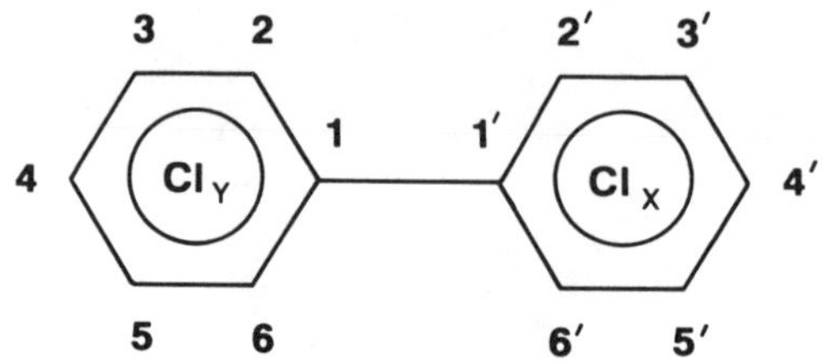

FIGURE 1. PCB configuration.

imately one half the world total. Monsanto produced PCBs labeled Aroclor 1221, 1232, 1016, 1242, 1248, 1254, 1260, 1262, and 1268, consisting of 21, 32, 41, 42, 48, 54, 60, 62, and 68% chlorine by weight, respectively.[21] Domestic sales of PCBs from 1929 to 1975 were 5.7×10^{11} g, of which 77% were used in electrical equipment.[7] From 1957 through 1971, PCBs with lower chlorine percentage by weight (Aroclors 1221, 1232, and 1242) accounted for 61% of sales (60% was 1242), while the heavier chlorinated Aroclors, led by 1254 with 15% and 1260 with 13%, accounted for the remainder. Monsanto voluntarily ceased production in 1977, and imports were terminated in 1980 in response to compliance with the TSCA which called for a ban on the use, manufacture, and distribution of PCBs by 1980. As of 1979, capacitors and transformers employing 3.4×10^{11} g of PCBs were estimated to be still in service.[7]

B. Health Effects

The potential for exposure to PCBs and PCB-contaminated substances was brought to the public's attention in 1966 when fish contaminated with PCBs were found in the Baltic Sea. Shortly thereafter, in 1968, PCB-contaminated rice oil (approximately 2000 mg of PCB per kilogram of oil) was ingested by more than 1500 members of a Japanese population.[2] An average 800 ml of the rice oil was ingested per person over a 3-month period, with daily intakes in the order of 200 μg/kg of body weight. In addition to 51 deaths (1968 to 1977), human health effects manifested after this accidental exposure in Japan included: (1) dark pigmentation of skin, mucous membranes, lips, and gums; (2) acne-like skin eruptions; (3) discharge from the eyes; (4) neurological disorders such as headache, numbness, and pain in the extremities; (5) possible hormonal disorders; (6) liver enlargment and stimulation of liver enzyme systems; (7) bone and dental disorders; and (8) reduced growth in boys (not detected in girls).[10] Because carcinogenic effects often take many years to appear, cancer risk conclusions have not been drawn from the Japanese data.[4]

Animal bioassay data support the PCB epidemiological information collected from industrial accidents and contamination of food supplies. Effects noted in animal species fed with PCB-contaminated foodstuffs (PCBs from 25 to 500 ppm) include: (1) fluid accumulation under the skin; (2) hair loss; (3) skin eruptions similar to acne; (4) increased formation and excretion of blood-forming compounds in the urine; (5) enlarged liver cells; (6) suppression of the immune response; (7) liver cancer; and (8) birth defects (noted in chicks only).[4]

C. Incineration Requirements

As discussed in Chapter 7, if incomplete combustion of PCBs occurs during incineration, both PCBs and the highly toxic dioxins and furans may be emitted to the atmosphere. Buser and Rappe[3] observe that PCBs pyrolized at 600°C yield furans estimated to be in the range of 0.1% up to several percent. Thus, when considering the risk of thermal disposal of PCBs, one must consider not only destruction and removal of PCBs, but destruction and removal of polychlorinated dibenzo-*p*-dioxins (PCDDs) and furans (PCDFs) as well.

To protect human health from the emissions associated with PCB incineration, the EPA has issued guidelines for PCB incineration. Table 1 lists the incinerator requirements for

Table 1
REQUIREMENTS FOR PCB INCINERATORS

Destruction efficiency	99.9999%
Temperature	1150°C
Residence time	2 sec
Excess oxygen	3%
CO	<100 ppm
HCl removal	99%
Particulates	State permit
PCB feed rate	Test burn

From U.S. Environmental Protection Agency, EPA-906/9-82-003, Dallas, TX, June 1981.

destruction of PCBs as set forth by the EPA. At present, even in light of potential health hazards, incineration appears to be the best commercial method of PCB destruction.[14]

III. TRIAL BURN STACK EMISSIONS

Three PCB trial burn series have been performed at land-based commercial incinerators to determine emission rates of PCBs. The first and second were conducted at the Energy Systems Company (ENSCO) incinerator in El Dorado, AR and at the Rollins Environmental Services facility in Deer Park, TX. The third PCB trial burn was conducted at the SCA Chemical Services incinerator in Chicago, IL.

A. Polychlorinated Biphenyl (PCB) Emissions

The ENSCO facility in Arkansas employs a rotary kiln incinerator, 34 feet long with an internal diameter of 7 feet. The refractory-lined kiln has a residence time of approximately 60 min for solids and 3.7 sec for gaseous material. The temperature inside the kiln varies from 1200 to 1500°F. Rollins Environmental Services at Deer Park consists of both a rotary kiln and liquid injection incinerator which discharge into a common afterburner. Temperatures up to 1500°C can be attained in the three burners, which have a combined residence time of 2 to 3 sec. The SCA incinerator is composed of a rotary kiln and secondary combustion chamber and operates at a maximum heat input rate of 120×10^6 Btu/h.

Table 2 summarizes PCB feed rates and destruction efficiencies for ENSCO, Rollins, and SCA. The input rates and destruction and removal efficiencies (DREs) for the ENSCO and Rollins facilities are average figures for the duration of the 1979 PCB trial burns; the SCA incinerator results are specific for each test run. The results indicate that all trial burns achieved greater than 99.9999% destruction efficiency (DE) for PCBs.

B. Dioxin and Furan Emissions

The current data base for PIC emissions during incineration of PCBs is very limited.[17] Because PCBs are composed of up to 60% chlorine by weight and since dioxin and furan emissions have been shown to be correlated to total chlorine input to the incinerator,[12] PIC emissions from PCB incinerators are suspected to contain dioxins and furans. The most complete data are from the three EPA-sponsored trial burns in Arkansas, Texas, and Illinois. The data on dioxin and furan emission rates obtained from these trial burns were surveyed in Chapter 7. The average total emission rate of dioxins and furans was 5.4E − 7% of the feed rate for the trial burn incinerators burning PCB waste (see Table 10). However, the results of these trial burns are subject to large uncertainty and could be in error by many

Table 2
PCB TRIAL BURN FEED RATES, DESTRUCTION EFFICIENCIES, AND EMISSION RATES[9,20]

Facility	Run no.	Input rate (g/min)	Destruction efficiency (%)	Stack emission rates (μg/sec)
ENSCO (1979)	Av	4975	99.99998	16.6
Rollins (1979)	Av	16764	99.99997	83.8
SCA				
	10	5599	99.99998	18.7
	11	9669	99.99999	16.1
	12	3164	99.99995	26.4
	13	17700	99.99999	29.5
	14	44500	99.99996	296.7

orders of magnitude.[17] Because stack gas samples were analyzed only for dioxins and furans, total PIC emissions resulting from incineration of PCBs are unknown.

IV. POLYCHLORINATED BIPHENYL (PCB) RISK ASSESSMENT

The purpose of this section is to determine the impact of incineration of waste containing PCBs on individuals living near the incinerator. Our assessment will focus on three hypothetical sites located in Arkansas, Texas, and Missouri. These locations were chosen because (1) they are representative of operating facilities which currently burn PCB waste and (2) they represent a broad range of population densities. Data for the Arkansas and Texas sites are obtained from the EPA;[17] data for the Missouri site are obtained from Holton et al.[8]

A. Air Dispersion

A waste feed rate of 56,000 t/year is assumed for the Arkansas and Texas facilities, 35% of which (19,600 t/year) is considered to be PCBs. A DRE of 99.9999% is used to calculate the PCB emission rate. Products of incomplete combustion (PIC) emission rates (dioxins and furans) used for the two sites are average rates derived from trial burns of PCB waste.[17] Holton et al.[8] assume a 135 × 10^6 Btu/h liquid injection incinerator located near Kansas City, MO, burning a waste stream composed of 18.54% PCBs. The waste feed rate at the Missouri site is 53,400 t/year. The same DRE, 99.9999%, is assumed for the exposure assessment.

To obtain stack emission rates for PCBs, waste throughput is multiplied by the fractional PCB representation in the waste stream and by 10^{-6} to account for the DRE chosen. To obtain dioxin and furan emission rates, waste throughput is multiplied by the factor 5.4E − 9, which represents dioxin and furan emission as a fraction of waste throughput (see Chapter 7, Table 10). Table 3 lists the data necessary for stack emission predictions at the three sites.

The Gaussian plume atmospheric transport model (ATM) is used together with the data in Table 3 to predict annual-average ground-level air concentrations of pollutants in the assessment areas.[17] Concentrations are determined at the center of each of 160 sector segments within 50 km of the incinerator stacks.

B. Exposure

The total population within 50 km of the Texas site is 2,500,000, the Arkansas assessment area population is 91,000, and the Missouri assessment area population is 70,483. Three

Table 3
DATA FOR STACK EMISSIONS PREDICTIONS[8,17]

Parameter	Arkansas	Texas	Missouri
Latitude	N 33 12 19	N 29 43 35	N 38 20 00
Longitude	W 92 37 45	W 95 05 40	W 94 20 00
Waste feed rate (mt/year)	56,000	56,000	53,400
DRE	99.9999	99.9999	99.9999
Stack height (m)	61.0	30.4	27.4
Stack radius (m)	1.45	0.90	1.04
Exit gas velocity (m/sec)	4.6	7.2	6.4
Exit gas temperature (K)	336	323	336.5
PCB emission rate (g/sec)	6.21E − 4	6.21E − 4	3.1E − 4
PIC emission rate (g/sec)	9.6E − 6	9.6E − 6	1.7E − 6

Table 4
EXPOSURES FROM PCB WASTE INCINERATION[17]

	Maximum individual exposure ($\mu g/m^3$)	Expected individual exposure ($\mu g/m^3$)	Population exposure (person — $\mu g/m^3$)
Arkansas	4.66E − 5	7.18E − 6	5.99E − 1
Texas	6.64E − 5	1.41E − 6	3.62E + 0
Missouri	5.88E − 5	4.26E − 6	2.96E + 0

measures of inhalation exposure are presented in Table 4: (1) maximum individual exposure, the highest pollutant concentration (micrograms per meter^3) at a *currently populated* receptor point; (2) expected (or average) individual exposure (micrograms per meter^3), and (3) total population exposure (person — micrograms per meter^3). These measures of inhalation exposure are obtained from the predicted annual average ground-level air concentrations and the population distributions around the incinerator sites.

C. Health Risk Estimates

Cancer risks from incineration of PCB-containing wastes are estimated using the following procedures:

1. Maximum individual exposure (micrograms per meter^3) and expected individual exposure (micrograms per meter^3) are determined from ground-level air concentrations of pollutants in the assessment area. Population exposure (person — micrograms per meter^3) is determined by multiplying ground-level atmospheric pollutant concentrations by the number of individuals reading in each sector segment and adding the results for all sector segments.
2. These three measures of exposure are multiplied by a conversion factor (which takes into account average human weight, breathing rate, and percent retention of inhaled chemical) to yield the maximum individual dose (milligrams per kilogram per day), average individual dose (milligrams per kilogram per day), and the population dose (person — milligrams per kilogram per day).
3. The dose estimates are multiplied by cancer risk factors (developed by the EPA) to obtain the primary estimates of cancer incidence: maximum individual risk (MIR), expected individual risk (EIR), and population risk (PR).

Table 5
CARCINOGENIC RISK FACTORS

Constituent	$(mg/kg/d)^{-1}$
PCB	4.34E + 0
2,3,7,8-TCDD	1.56E + 5

From U.S. Environmental Protection Agency, EPA-600/8-83-014F, Environmental Criteria and Assessment Office, Research Triangle Park, N.C., August 1984.

Table 6
SUMMARY OF PCB CANCER RISK ESTIMATES FOR THREE SITES[8,17]

	Arkansas	Texas	Missouri
Population risk	5.5E − 4	3.3E − 3	2.7E − 3
Maximum individual risk	4.2E − 8	6.1E − 8	5.4E − 8
Expected individual risk	6.6E − 9	1.3E − 9	3.9E − 9

Table 5 lists the carcinogen potency factors[18] used in the calculations, and Table 6 summarizes the categories of risk estimates calculated for the three sites: (1) excess cancer incidence for the entire assessment area population (expected excess cancers in the exposed population during 70 years); (2) maximum risk to the individual (the probability that an individual will develop cancer after 70 years of continuous exposure at maximum concentrations); and (3) expected risk to the individual (the probability that an individual will develop cancer after 70 years of continuous exposure at the average predicted atmospheric concentration of PCBs around the incinerator sites).

D. Follow-Up

In the summer of 1981, residents of Deer Park, TX, the actual site of our hypothetical Texas incinerator, complained of skin rashes they believed to be caused by exposure to atmospheric chemicals. An investigation was consequently conducted by the Texas Air Control Board (TACB) laboratory.[15] Samples of area air, soil, water, and vegetation were analyzed for contamination by ten metals, nine organic compounds (including PCBs), and fluoride, which were the chemicals isolated as possible causative agents of skin rashes. In the four testing phases performed by TACB, none of the chemicals were detected in the environment at levels significant enough to cause adverse health effects such as skin rashes.[15]

V. PRODUCTS OF INCOMPLETE COMBUSTION (PIC) RISK ASSESSMENT

As pointed out in Chapter 7, data on PIC emission rates from PCB incinerators are limited. The only data available are dioxin/furan emission rates from three EPA trial burn facilities burning PCB waste. Health risk estimates for the three trial burn facilities were presented in Chapter 7. These risk estimates indicate risk to the maximally exposed individual on the order of 1×10^{-5}; however, these estimates were made using the assumption that all dioxin/furan emissions from the incinerators are as toxic as 2,3,7,8-tetrachlorodibenzo-*p*-dioxin (2,3,7,8-TCDD). It is known that this is not the case. In order to obtain a more realistic

Table 7
RELATIVE POTENCY OF DIOXIN AND FURAN HOMOLOGUES[19]

PCDDs

Homologue	Acronym	Relative potency
Monochlorodibenzo-*p*-dioxin	MCDD	0
Dichlorodibenzo-*p*-dioxin	DCDD	0
Trichlorodibenzo-*p*-dioxin	TrCDD	0
Tetrachlorodibenzo-*p*-dioxin	TCDD	1
Pentachlorodibenzo-*p*-dioxin	PeCDD	0.5
Hexachlorodibenzo-*p*-dioxin	HxCDD	0.04
Heptachlorodibenzo-*p*-dioxin	HpCDD	0.001
Octachlorodibenzo-*p*-dioxin	OCDD	0

PCDFs

Homologue	Acronym	Relative potency
Monochlorodibenzofuran	MCDF	0
Dichlorodibenzofuran	DCDF	0
Trichlorodibenzofuran	TrCDF	0
Tetrachlorodibenzofuran	TCDF	0.1
Pentachlorodibenzofuran	PeCDF	0.1
Hexachlorodibenzofuran	HxCDF	0.01
Heptachlorodibenzofuran	HpCDF	0.001
Octachlorodibenzofuran	OCDF	0

estimate of health effects resulting from exposure to dioxin/furan emissions at PCB incinerators, the toxicity equivalence factor approach outlined in Chapter 7 (the section on 2,3,7,8-TCDD toxicity equivalence factors) will be applied.

A. Toxic Equivalents

Table 7 presents the relative potency factor of the various homologues of PCDD/PCDF.[19]

A risk assessment of dioxin and furan emissions, using the toxicity equivalence factor approach, involves five steps (see the section on 2,3,7,8-TCDD toxicity equivalence factors in Chapter 7), one of which (step three) is the determination of the toxic equivalent emission rate. Table 8 presents the toxic equivalent emission rate of dioxins and furans, as a percentage of the waste feed rate, for an incinerator burning PCB waste. This table is based on dioxin and furan emission rates given in Table 10 of Chapter 7 and the relative toxicity data in Table 7. The average total dioxin and furan emission rate over all the trial burn data, expressed in toxic equivalents of 2,3,7,8-TCDD (as a percentage of the waste feed rate), is 3.9×10^{-8}.

B. Health Risk Estimates

Cancer risk estimates from dioxin and furan emissions associated with incineration of PCB wastes are estimated using the following procedure:

1. The toxic equivalent emission rate is determined by multiplying the waste throughput by the factor 3.9×10^{-8}, which represents the 2,3,7,8-TCDD toxic equivalent emissions as a fraction of waste throughput (see Table 8).

Table 8
TOXIC EQUIVALENT DIOXIN AND FURAN EMISSIONS (% OF WASTE FEED) AT TRIAL BURN INCINERATORS BURNING PCB WASTE

Homologue	Average emission[a] rate (%)	Relative[b] potency	Toxic equivalent emission rate (%)
TCDD	3.4E − 8	1	3.4E − 8
TCDF	1.1E − 8	0.1	1.1E − 9
PeCDF	4.0E − 8	0.1	4.0E − 9
HxCDF	2.5E − 8	0.01	2.5E − 10
HpCDF	1.3E − 9	0.001	1.3E − 12
Total (%)			3.9E − 8

[a] From References 9, 17, and 20.
[b] From Reference 19.

Table 9
SUMMARY OF DIOXIN AND FURAN CANCER RISK ESTIMATES FOR THREE SITES[8,17]

	Arkansas	Texas	Missouri
Population risk	8.6E − 3	5.1E − 2	4.1E − 2
Maximum individual risk	6.4E − 7	9.3E − 7	8.6E − 7
Expected individual risk	9.3E − 8	2.0E − 8	6.0E − 8

2. Maximum individual exposure (micrograms per meter^3) and expected individual exposure (micrograms per meter^3) are determined from ground-level air concentrations of pollutants in the assessment area. Population exposure (person — micrograms per meter^3) is determined by multiplying ground-level atmospheric pollutant concentrations by the number of individuals reading in each sector segment and adding the results for all sector segments.
3. These three measures of exposure are multiplied by a conversion factor (which takes into account average human weight, breathing rate, and percent retention of inhaled chemical) to yield the maximum individual dose (milligrams per kilogram per day), average individual dose (milligrams per kilogram per day), and the population dose (person — milligrams per kilogram per day).
4. The dose estimates are multiplied by the cancer risk factor for 2,3,7,8-TCDD (developed by the EPA) to obtain the primary estimates of cancer incidence: MIR, EIR, and PR.

Table 9 summarizes the categories of risk estimates calculated for the three sites.

VI. SUMMARY OF INHALATION RISK

The incidence of cancer from inhalation exposure at the three sites is remarkably similar. The MIRs for inhalation exposure at the Missouri site to PCBs and dioxins/furans are 5.4E − 8 and 1.9E − 6, respectively.[8] That is, an individual has a 5.4 in 100 million chance of developing cancer after 70 years of continuous exposure to PCB emissions at the site, and almost a two in one million probability of developing cancer after lifetime exposure to dioxin

Table 10
COMPARISON OF MIR ESTIMATES AT THREE TRIAL BURN SITES AND THREE HYPOTHETICAL SITES[8,9,17,20]

	Arkansas	Texas	Missouri
PCBs	4.2E − 8	6.1E − 8	5.4E − 8
Dioxins/furans	1.4E − 6	2.1E − 6	1.9E − 6

and furan emissions. Texas and Arkansas MIR estimates for lifetime PCB exposures are 6.1 chances in 100 million and 4.2 chances in 100 million. Thus, MIRs at all three facilities are in the same order of magnitude for PCBs, PCDDs, and PCDFs, with the exception of a smaller risk from PCDDs and PCDFs at the Texas facility; EIRs for all organics at the three incinerators are also in the same order of magnitude.

Table 10 compares the inhalation risk estimates for the three hypothetical PCB incineration facilities. The estimates for all facilities are within the range of 10^{-6} to 10^{-8}.

VII. CONCLUSION

At this writing, only three commercial, land-based, operating incineration facilities (ENSCO, Rollins, and SCA) have EPA permits to burn PCBs and PCB-contaminated waste. State health officials in Maryland, however, issued a permit in September 1985 for the Army to burn PCBs in an incinerator at the Aberdeen Proving Ground.[16] The burning of 3300 gallons of PCBs per year is permitted. The Army has agreed that hazardous materials will not be brought to the base from the outside for destruction. A risk assessment for this facility is not currently available.

REFERENCES

1. **Allen, J. R. and Norback, D. H.,** Pathobiological Responses of primates to polychlorinated biphenyl exposure, in Proceedings, National Conference of Polychlorinated Biphenyls, Ayer, F. A., Ed., EPA 560/6-75-004, U.S. Environmental Protection Agency, Research Triangle Park, NC, 1976, 43.
2. **Ayer, F. A.,** National Conference on Polychlorinated Biphenyls (November 19 to 21, 1975, Chicago, Illinois), EPA 560/6-75-004, U.S. Environmental Protection Agency, Washington, D.C., March 1976.
3. **Buser, H. R. and Rappe, C.,** Formation of polychlorinated dibenzofurans (PCDFs) from the pyrolysis of individual PCB isomers, *Chemosphere,* 8, 157, 1979.
4. Disposal of Polychlorinated Biphenyls in California, LE-80-007, California Air Resources Board, September 1980.
5. Code of Federal Regulations 40 CFR, 761.20, 1985.
6. **D'Itri, F. M. and Kamrin, M. A., Eds.,** *PCBs: Human and Environmental Hazards,* Butterworth Publishers, Woburn, MA, 1983, 7.
7. **Golden, D. M., Project Manager,** *Disposal of Polychlorinated Biphenyls (PCBs) and PCB-contaminated Materials,* Vol. 1, FP-1207, 1263-1, Electric Power Research Institute, Palo Alto, CA, 1979.
8. **Holton, G. A., Travis, C. C., and Etnier, E. L.,** A comparison of human exposures to PCB emissions from oceanic and terrestrial incineration, *Hazardous Waste Hazardous Mater.,* 2, 453, 1985.
9. **Chas. T. Main, Inc.,** Health Risk Assessment of the Incineration of Polychlorinated Biphenyls, prepared for SCA Chemical Services, Inc. by Chas. T. Main, Boston, MA, 1983.
10. **Nagayama, J., Masuda, Y., and Kuratsune, M.,** Chlorinated dibenzofurans in kanechlors and rice oils used by patients with Yusho, *Fukuoka Acta Med.,* 66(10), 593, October 1975.
11. National Academy of Sciences, Polychlorinated Biphenyls, National Research Council/National Academy of Science, Washington, D.C., 1979.

12. **Oberg, T., Aittola, J. P., and Bergstrom, J. G. T.,** Chlorinated aromatics from the combustion of hazardous waste, *Chemosphere,* 14(2), 215, 1985.
13. **Safe, S.,** Overview of analytical identification and spectroscopic properties, in Proceedings, National Conference on Polychlorinated Biphenyls, Ayer, F. A., Ed., EPA 560/6-75-004, U.S. Environmental Protection Agency, Research Triangle Park, NC, 1976, 94.
14. Science Advisory Board, Environmental Effects, Transport and Fate Committee, Report on the Incineration of Liquid Hazardous Wastes, U.S. Environmental Protection Agency, Washington, D.C., April, 1985.
15. Texas Air Control Board, Clear Blue, Vol. 9, No. 2, Texas Air Control Board, Austin, TX, 1982.
16. The Aegis, Hartford, Co., Maryland, September 26, 1985, p. A-1. The national incinerator testing and evaluation program, presented by the Environmental Protection Service, Urban Activities Division, Quebec, at the 4th Int. Symp. Chlorinated Dioxins and Related Compounds, Ottawa, Canada, October 16 to 18, 1984.
17. Assessment of Incineration as a Treatment Method for Liquid Organic Hazardous Wastes, U.S. Environmental Protection Agency, Washington, D.C., March, 1985.
18. Health Assessment Document for Chromium, EPA-600/8-83-014F, Environmental Criteria and Assessment Office, U.S. Environmental Protection Agency, Research Triangle Park, NC, August 1984.
19. Interim Procedures for Estimating Risks Associated with Exposures to Mixtures of Chlorinated Dibenzo-p- *Dioxins and Dibenzofurans (CDDs and CDFs),* U.S. Environmental Protection Agency, Washington, D.C., 1986.
20. PCB Disposal by Thermal Destruction, EPA-906/9-82-003, U.S. Environmental Protection Agency, Dallas, TX, June 1981.
21. Polychlorinated biphenyls (PCBs), toxic pollutant effluent standards, Federal Register, 42(22), 6532, U.S. Environmental Protection Agency, February 2, 1977.

Chapter 9

FOOD CHAIN ANALYSIS

I. INTRODUCTION

The inhalation pathway is the primary focus of most studies of population exposures resulting from hazardous waste incineration.[10,19,21] Although little attention is given to the terrestrial food chain pathway in previous incineration studies, investigations of synthetic fuel production[26] and coal-fired power plants[15] show that the food chain can be an important contributor to total population exposure. The purpose of this chapter is to determine the impact of hazardous waste incinerators on food crops grown in the vicinity of the facilities. The potential pathways for contamination of vegetation are shown in Figure 1.

Our assessment will focus on pesticide-related wastes. Pesticide-related chemicals are chosen because of their inherent toxicity and tendency to bioaccumulate in biota. Table 1 lists the ten most prevalent pesticide-related chemicals found in hazardous waste streams currently being incinerated. We will consider the impact of long-term incineration of a single generic waste stream containing all ten of these pesticide-related chemicals. Note that it is highly unlikely that all ten of these chemicals would ever be found in a single waste stream.

For this assessment, identical 150×10^6 Btu/h (4400 kw) liquid injection incinerator facilities (burning pesticide-related waste) are assumed to be sited in three different U.S. locations. The three locations studied are a southern California site (S-1), a northern Midwest site (S-2) and a central Midwest site (S-3). These sites are in areas that lead the nation in production of leafy vegetables, milk, and beef, respectively, and are chosen to estimate possible worst case contaminations of these foodstuffs.

II. STACK EMISSION SOURCE TERMS

As discussed in previous chapters, the rate of release (mass emissions per unit time) of specific chemicals in stack emissions is controlled by three facility variables: (1) waste throughput, (2) chemical concentration in the waste stream, and (3) destruction and removal efficiency (DRE). Waste throughput is estimated to be 2.76×10^{10} g/year. The standard DRE of 99.99% is assumed for this assessment. To obtain stack emission rates, throughput is multiplied by the fractional representation of each pesticide in the waste stream and by 10^{-4} to account for the 99.99% DRE. The fractional representation of each pesticide in the generic waste stream is taken to be equal to the average fractional representation of that pesticide in hazardous waste streams currently being incinerated in the U.S.[24] Predicted annual stack emission rates (grams per year) of the ten pesticide-related chemicals are given in Table 2.

III. POPULATION DISTRIBUTION

The number of persons living within a 100 km radius of each facility is obtained from 1980 census data. The central Midwest site (S-3) is the most populous with 1.46 million people, followed by the northern Midwest site (S-2) with 0.45 million people, and the southern California site (S-1) with 0.20 million people. Cumulative population distributions for the three sites are given in Table 3.

IV. ATMOSPHERIC DISPERSION MODEL

Estimates of average ground-level air concentrations for each pollutant are made using the atmospheric transport model (ATM), a Gaussian plume dispersion model developed at

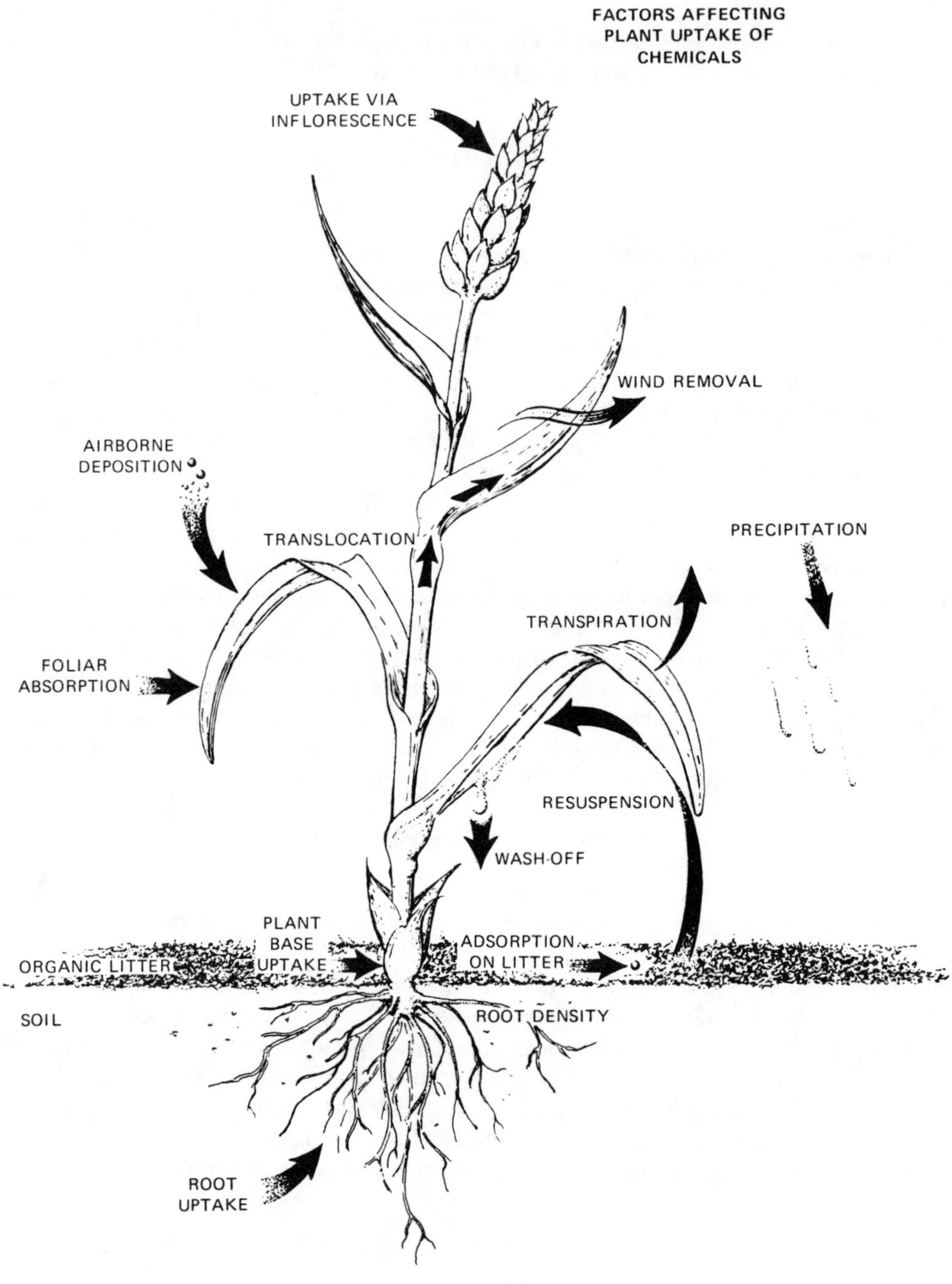

FIGURE 1. Illustrated potential pathways for contamination of vegetation.

Oak Ridge National Laboratory.[17] The ATM is selected because, in addition to estimating air concentrations, it specifically calculates the deposition rate of each pollutant onto the ground, a value necessary as input into the terrestrial food chain model.

Input parameters necessary to run the ATM include stack parameters for the hazardous waste incinerators and site-specific meterological data. These input parameters are summarized in Tables 4 and 5.

Table 1
TEN MOST PREVALENT PESTICIDE-RELATED CHEMICALS FOUND IN INCINERATION WASTE STREAMS[24]

Constituent	Amount incinerated (t/year)
1,1,1,2-Tetrachloroethane	9.57E4
Chloroform	9.13E4
Ethylene dichloride	7.85E4
Hexachlorobutadiene	6.79E4
1,1,2,2-Tetrachloroethane	2.10E4
Hexachloroethane	1.45E4
Carbon tetrachloride	1.35E4
Hexachlorobenzene	1.23E4
Trichloropropane	1.20E4
Trichloroethylene	1.15E4

Table 2
STACK EMISSION RATES OF THE PESTICIDE-RELATED WASTE (99.99% DRE)

Constituent	Fraction of waste stream[a] (%)	Emission rate (g/year)
1,1,1,2-Tetrachloroethane	6.6	1.95E4
Chloroform	6.1	1.80E4
Ethylene dichloride	19.0	5.61E4
Hexachlorobutadiene	9.2	2.72E4
1,1,2,2-Tetrachloroethane	2.4	7.08E3
Hexachloroethane	2.3	6.79E3
Carbon tetrachloride	0.9	2.66E3
Hexachlorobenzene	1.7	5.02E3
Trichloropropane	11.7	3.45E4
Trichloroethylene	2.7	7.97E3

[a] Based on the average fractional representation of each pesticide in hazardous waste streams currently being incinerated in the U.S.[24]

Table 3
CUMULATIVE POPULATION AT THREE INCINERATOR SITES

Distance (km)	S-1	S-2	S-3
10.0	0	7,432	2,528
20.0	992	76,571	10,474
30.0	4,984	108,979	18,777
40.0	8,585	128,609	42,134
50.0	28,827	196,307	70,483
60.0	55,307	263,092	143,535
70.0	78,809	317,157	309,225
80.0	110,132	349,963	724,409
90.0	141,703	395,472	1,168,924
100.0	191,007	448,182	1,455,344

Table 4
STACK PARAMETERS[21]

Height above ground (m)	27.43
Gas temperature (°K)	366.50
Release velocity (m/sec)	6.40
Stack radius (m)	1.04

Table 5
METEOROLOGICAL PARAMETERS INPUT INTO ATM[21]

Parameter	S-1	S-2	S-3
Annual average air temperature (°K)	295.8[a]	280.2[b]	286.3[c]
Annual average rainfall (in./y)	3.16[a]	28.72[b]	35.54[c]
Number of days with 0.01 in. or less of rainfall	15[d]	120[e]	107[f]
Fraction of time that it rains[g]	0.0103	0.0822	0.0733
Annual average rainfall intensity[h] (in./h)	3.51	3.99	5.54
Average morning mixing-layer height (m)	536	458	413
Average afternoon mixing-layer height (m)	1210	1213	1349

[a] Average for three weather stations: Blythe/Riverside and Imperial, CA (1931—1960) and Yuma, AZ (1941—1970).
[b] Average for two weather stations: Eau Claire (1931—1960) and Green Bay, WI (1941—1970).
[c] Average of 1931—1960 and 1941—1970 data for Kansas City, MO.
[d] Average for station at Yuma, AZ (1941—1970).
[e] Average for station at Green Bay, WI (1941—1970).
[f] Average for station at Kansas City, MO (1941—1970).
[g] Calculated as 0.25 × (number of days with ≥ 0.01 in. of rainfall per 365-d year).
[h] Calculated as annual average rainfall (in./year)/(8760 h/year × fraction of time that it rains).

V. INGESTION EXPOSURES

Human exposures via the food chain pathway from incineration of pesticide-related waste chemicals are estimated using the following procedure:

1. Ground-level atmospheric concentrations of pesticides are determined within a 100 km radius of each facility using stack emission rates (Table 2) and the ATM. Concentrations are determined at the center of each of 160 sector segments (as in Chapter 5, Figure 2).
2. Deposition rates of pesticides onto soil and terrestrial food crops are determined at the center of each of the 160 sector segments.
3. Concentrations of the pesticides in various food and forage crops are determined using a food chain uptake model (a description of this model is in the appendix to this chapter). The food chain model accounts for two components in the food crop uptake process (see Figure 2): (a) a soil uptake component in which the pollutant is first deposited on the soil and then taken up by the roots of the plant and translocated to the edible portion of the vegetation and (b) an atmospheric deposition component in which the pollutant is deposited directly onto the plant. Actual crop production data in each of the 160 sector segments are used to determine the type and quantity of crops exposed to the pesticides (see Figure 3).
4. Once pollutant concentrations in foodstuffs are estimated, individual intakes of each chemical are calculated based on dietary intake of specific foods. Dietary components considered include vegetables, grains, dairy products, and meat (see the appendix for more detail).

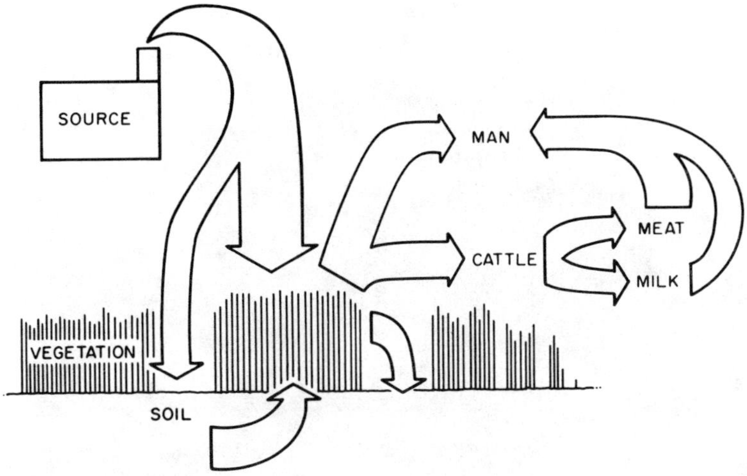

FIGURE 2. Atomospheric deposition and uptake through the food chain.

Table 6 lists minimum, maximum, and expected individual exposures (doses) from the food chain pathway assuming 95% absorption through the gastrointestinal tract. Expected (average) individual ingestion intakes (micrograms per year) at all three sites are very similar, differing at most by a factor of approximately four. This indicates that local meteorology and variability in food crop production and yield in the vicinity of incinerators across the U.S. appear to have little impact on exposure from the food chain pathway. Thus, the expected ingestion intakes presented in Table 6 are probably representative of intakes around any incinerator located in the U.S. (given the scenario considered in this chapter).

Table 7 lists the percent of the contribution of each food item to total dietary intake of hexachlorobenzene by an individual at Site S-2. The beef pathway is the most important, contributing 49% to total intake; grain and leafy vegetables contribute 19% and 13%, respectively. Variability in individual dietary habits (for example, a vegetarian diet) could lead to variability in these percentages.

Not all food produced at the three sites is consumed locally. The most productive site, S-1, produces approximately 20 times more food (kilograms per year) than can be consumed by the local population. Site S-2 produces about 16 times more than its population can consume, while site S-3 only produces a fraction (0.84) of its local needs. The excess food production at sites S-1 and S-2 may be exported to other parts of the U.S., potentially resulting in ingestion exposures to populations removed from these two sites. However, since most of the diet of an individual in other parts of the U.S. would consist of uncontaminated food, these exposures can be expected to be much smaller than those of the maximum intakes (micrograms per year) given in Table 6.

VI. INHALATION EXPOSURE

To place the expected individual food chain exposures in perspective, human inhalation exposures from incineration of the same pesticide-related waste are estimated at the three sites (S-1, S-2, and S-3). The procedure, similar to that used for estimating exposures from the food chain pathway, is as follows:

1. Atmospheric concentrations of each pesticide are determined within a 100 km radius of each incinerator facility using stack emission rates (Table 2) and the ATM. Concentrations are determined at the center of each of 160 sector segments (as in Chapter 5, Figure 2).

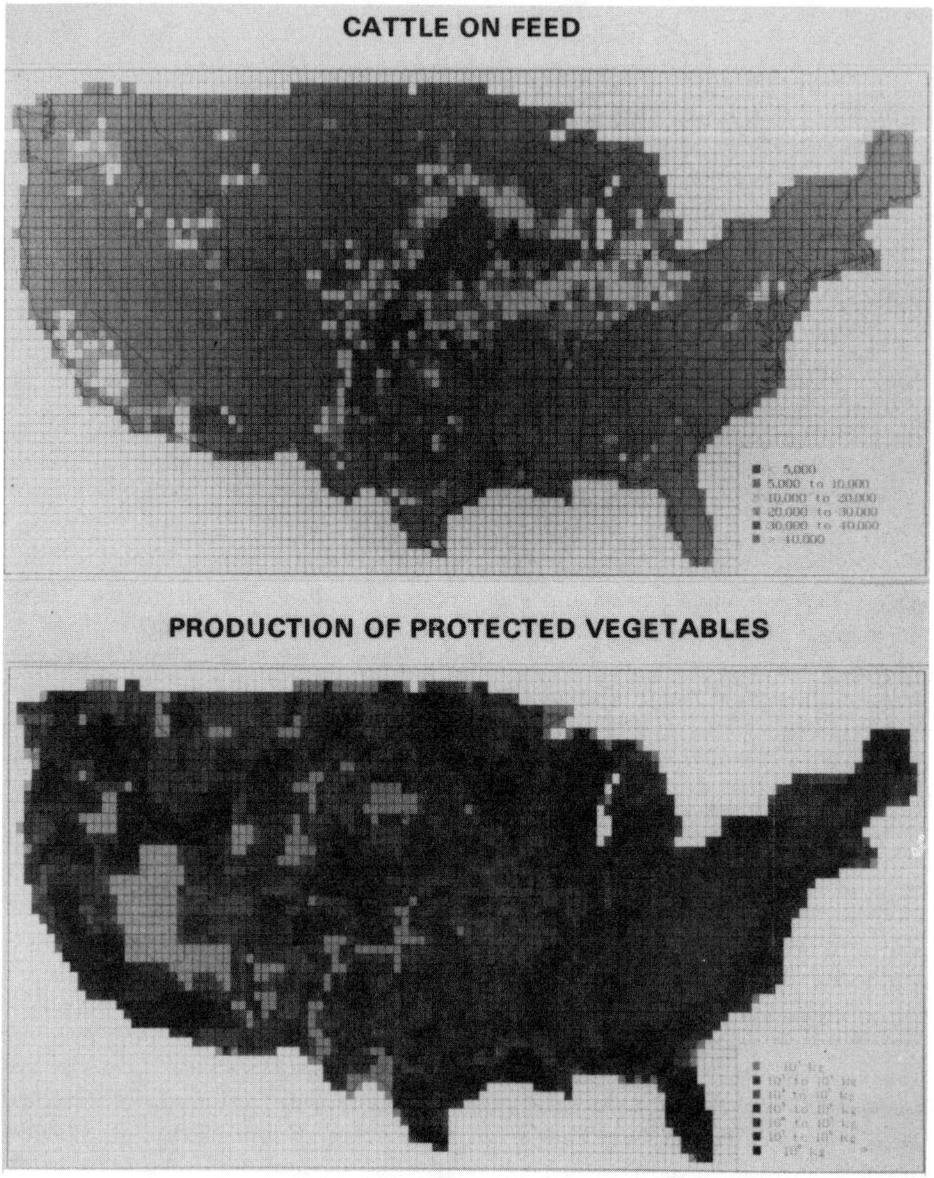

FIGURE 3. Samples of agricultural data used to determine crop and beef contamination.

2. Inhalation exposures in a sector segment are calculated by multiplying the average ground-level air concentration in the sector segment by the number of persons located in that segment. Pollutant intake calculations employ a breathing rate of 8322 m^3/year and a lung absorption factor of 0.65.

Table 8 lists minimum, maximum, and expected individual intakes from the inhalation pathway for each of the three sites. Although total population size and distribution vary between the three sites, expected average individual inhalation intakes of each chemical at all three sites are very similar, varying by a factor of only 2.6. This indicates that local meteorology and population distributions have little impact on average inhalation intakes. Thus, the expected inhalation intakes presented (Table 8) are probably representative of intakes around any incinerator in the U.S. (given the scenario considered in this chapter).

Table 6
INGESTION INTAKE[a] (μg/YEAR) AT THREE SITES[22]

Pollutant	Site	Minimum[b]	Expected[c]	Maximum[d]
1,1,1,2-Tetrachloroethane	S-1	2.40E − 3	2.85E − 2	6.74E − 1
	S-2	2.55E − 2	1.24E − 1	2.52E + 0
	S-3	1.17E − 2	1.09E − 1	3.94E + 0
Chloroform	S-1	2.82E − 3	3.83E − 3	8.35E − 1
	S-2	3.09E − 2	1.46E − 1	3.11E + 0
	S-3	1.37E − 2	1.09E − 1	3.94E + 0
Ethylene dichloride	S-1	3.91E − 4	4.25E − 3	6.23E − 2
	S-2	3.46E − 3	1.77E − 2	2.74E − 1
	S-3	1.65E − 3	1.18E − 2	4.29E − 1
Hexachlorobutadiene	S-1	9.32E − 5	8.04E − 4	1.04E − 2
	S-2	6.50E − 4	3.74E − 3	4.95E − 2
	S-3	2.81E − 4	1.95E − 3	7.03E − 2
1,1,2,2-Tetrachloroethane	S-1	4.17E − 4	4.49E − 3	6.65E − 2
	S-2	4.42E − 3	1.88E − 2	3.21E − 1
	S-3	1.73E − 3	1.31E − 2	4.11E − 1
Hexachloroethane	S-1	1.33E − 3	1.38E − 2	2.79E − 1
	S-2	1.29E − 2	6.04E − 2	9.61E − 1
	S-3	4.86E − 3	3.83E − 2	1.14E + 0
Carbon Tetrachloride	S-1	2.72E − 3	4.30E − 2	1.62E + 0
	S-2	1.14E − 2	5.08E − 2	1.28E + 0
	S-3	4.10E − 3	2.91E − 2	1.23E + 0
Hexachlorobenzene	S-1	9.70E − 4	8.07E − 3	1.12E − 1
	S-2	8.19E − 3	4.48E − 2	8.00E − 1
	S-3	4.71E − 3	2.68E − 2	5.15E − 1
Trichloropropane	S-1	3.62E − 4	4.00E − 3	6.29E − 2
	S-2	3.49E − 3	1.60E − 2	2.22E − 1
	S-3	1.31E − 3	9.21E − 3	3.14E − 1
Trichloroethylene	S-1	1.12E − 4	1.41E − 3	1.94E − 2
	S-2	1.04E − 3	5.86E − 3	7.48E − 2
	S-3	4.21E − 4	2.82E − 3	4.76E − 2

[a] Stack emissions calculated from expected throughput and pollutant concentrations in waste for the 150 × 10^6 Btu/h liquid injection incinerator (99.99% DRE). Absorption of 95% through the gastrointestinal tract is assumed; standard diet assumed (see Appendix to this chapter).

[b] Minimum sector segment receptor uptake.

[c] Average uptake for the assessment region.

[d] Maximum sector segment receptor uptake.

Table 7
HEXACHLOROBENZENE INTAKE FOR AN INDIVIDUAL AT SITE S-2[22]

Food item	Contribution (%)
Leafy vegetables	13
Protected vegetables	4
Exposed vegetables	8
Grains	19
Beef	49
Milk	8
Total	100

Table 8
INHALATION INTAKE[a] (µg/YEAR) AT THREE SITES[21,22]

Pollutant	Site	Minimum[b]	Expected[c]	Maximum[d]
1,1,1,2-Tetrachloroethane	S-1	2.18E − 2	7.74E − 2	3.37E − 1
	S-2	1.77E − 2	1.16E − 1	5.27E − 1
	S-3	1.95E − 2	4.46E − 2	6.37E − 1
Chloroform	S-1	2.03E − 2	7.15E − 2	3.12E − 2
	S-2	1.63E − 2	1.10E − 1	4.87E − 1
	S-3	1.80E − 2	4.13E − 2	5.88E − 1
Ethylene dichloride	S-1	6.30E − 2	2.23E − 1	9.71E − 1
	S-2	5.11E − 2	3.34E − 1	1.52E + 0
	S-3	5.63E − 2	1.29E − 1	1.83E + 0
Hexachlorobutadiene	S-1	3.04E − 2	1.08E − 1	4.69E − 1
	S-2	2.47E − 2	1.61E − 1	7.35E − 1
	S-3	2.73E − 2	6.22E − 2	8.87E − 1
1,1,2,2-Tetrachloroethane	S-1	7.96E − 3	2.82E − 2	1.22E − 2
	S-2	6.45E − 3	4.23E − 2	1.92E − 1
	S-3	7.11E − 3	1.62E − 2	2.31E − 1
Hexachloroethane	S-1	7.56E − 3	2.69E − 2	1.18E − 1
	S-2	6.17E − 3	4.04E − 2	1.84E − 1
	S-3	6.79E − 3	1.55E − 2	2.21E − 1
Carbon tetrachloride	S-1	2.98E − 3	1.05E − 2	4.59E − 2
	S-2	2.42E − 3	1.58E − 2	7.19E − 2
	S-3	2.66E − 3	6.08E − 3	8.67E − 2
Hexachlorobenzene	S-1	5.63E − 3	1.99E − 2	8.67E − 2
	S-2	4.56E − 4	2.99E − 2	1.34E − 1
	S-3	5.02E − 3	1.15E − 2	1.63E − 1
Trichloropropane	S-1	3.89E − 2	1.38E − 1	6.00E − 1
	S-2	3.15E − 2	2.06E − 1	9.42E − 1
	S-3	3.47E − 2	7.94E − 2	1.13E + 0
Trichloroethylene	S-1	8.95E − 3	3.16E − 2	1.38E − 1
	S-2	7.25E − 3	4.75E − 2	2.16E − 1
	S-3	8.00E − 3	1.83E − 2	2.60E − 1

[a] Stack emissions calculated from expected throughput and pollutant concentrations in waste for the 150 × 10^6 Btu/h liquid injection incinerator (99.99% DRE). Absorption of 95% through the gastrointestinal tract is assumed; standard diet assumed (see Appendix to this chapter).
[b] Minimum sector segment receptor uptake.
[c] Average uptake for the assessment region.
[d] Maximum sector segment receptor uptake.

VII. CONCLUSION

Comparison of the average individual inhalation and ingestion intake (micrograms per year) from incineration of pesticide-related waste at site S-1 is shown in Table 9. The intake values are similar for several of the chemicals. Table 10 presents the percent of the contribution of the food chain pathway to total average individual dose (inhalation plus food chain) for the three sites studied. The contribution of the food chain to total dose is chemical specific, depending on the bioaccumulation factors and the atmospheric degradation rates (see the appendix). The percent of the contribution of the ingestion dose to total dose ranges from 0.7% for hexachlorobenzene to 83% for carbon tetrachloride. Differences between sites result primarily from site-specific differences in productivity rates for the various food items.

Although there is a large uncertainty associated with ingestion exposure calculations, the data herein suggest that the ingestion pathway is of comparable import to the direct inhalation

Table 9
COMPARISON OF AVERAGE INDIVIDUAL INHALATION AND INGESTION INTAKE (μg/YEAR) FROM INCINERATION OF PESTICIDE-RELATED WASTE AT SITE S-1[22]

Pollutant	Inhalation (μg/year)	Ingestion (μg/year)
1,1,1,2-Tetrachloroethane	7.74E − 2	2.85E − 2
Chloroform	7.15E − 2	3.83E − 3
Ethylene dichloride	2.23E − 1	4.25E − 3
Hexachlorobutadiene	1.08E − 1	8.04E − 4
1,1,2,2-Tetrachloroethane	2.82E − 2	4.49E − 3
Hexachloroethane	2.69E − 2	1.38E − 2
Carbon tetrachloride	1.05E − 2	4.30E − 2
Hexachlorobenzene	1.99E − 2	8.07E − 3
Trichloropropane	1.38E − 1	4.00E − 3
Trichloroethylene	3.16E − 2	1.41E − 3

Table 10
PERCENT CONTRIBUTION OF FOOD CHAIN PATHWAY TO TOTAL AVERAGE INDIVIDUAL DOSE (INHALATION PLUS FOOD CHAIN)[22]

Pollutant	S-1	S-2	S-3
1,1,1,2-Tetrachloroethane	27	60	65
Chloroform	5	57	73
Ethylene dichloride	2	5	8
Hexachlorobutadiene	0.7	2	3
1,1,2,2-Tetrachloroethane	14	31	35
Hexachloroethane	34	60	71
Carbon tetrachloride	80	76	83
Hexachlorobenzene	29	60	70
Trichloropropane	3	7	10
Trichloroethylene	4	11	13

pathway. Inhalation exposures related to principal organic hazardous constituent (POHC) emissions from hazardous waste incineration are shown in Chapter 5 to be small in terms of cancer risk. Thus, since food chain and inhalation exposures from incineration of pesticide-related waste are comparable, it is suggested that food chain exposures resulting from POHCs released during incineration are not likely to be significant. (This assumes, of course, that cancer potency values for food ingestion are not significantly higher than for direct inhalation.)

Conversely, metal emissions from incinerators may have potentially significant health effects. Since the impact of metal emissions on food crops grown in the vicinity of incineration facilities is not analyzed in this book (and has not been analyzed by others as of this writing), this topic should be the focus of a future study.

I. APPENDIX

A. Food Chain Methodology

After contaminants are released by incinerators, they may enter the terrestrial food chain and pose health risks to man via ingestion of contaminated foods. Modeling the transport of organic contaminants through the food chain is a relatively recent endeavor; only a few food chain models are available in the literature.[4,11,20,22] These models consider human exposure pathways involving agricultural produce, beef, and milk. The conceptual framework on which the methodology herein is based is summarized in Figure 4. The transport pathways considered for plants are (1) direct deposition onto edible plant surfaces for two classes of vegetables and all feed crop categories and (2) root uptake of the contaminant from soil. Details of the methodology follow.

The methodology used to predict the uptake of organic chemicals by food and forage crops is given by the general equation:

$$C_p = d(A_d + U_s)$$

where C_p = concentration in plants (μg/kg); d = deposition rate (μg/m² s/kg); A_d = atmospheric deposition component (m² s/kg); and U_s = uptake from soil component (m² s/kg).

This equation is simply a statement that the concentration of a pollutant in plants is the result of one or both of the following processes: (1) an atmospheric deposition component in which the pollutant is deposited directly on the plant or (2) a soil uptake component in which the pollutant is first deposited on the soil and then taken up by the roots of the plant and translocated to the edible portion of the vegetation.

The atmospheric deposition component is given by:

$$A_d = \frac{r[1 - \exp(-\lambda_{al}\, t_e)]}{Y\, \lambda_{al}}$$

where r = interception fraction for the edible portion of the plant (unitless); λ_{al} = atmospheric loss constant (s^{-1}); t_e = time of exposure of edible plant parts to atmospheric deposition(s); and Y = yield or standing crop biomass of the edible portion of the plant (kg/m²).

The atmospheric loss constant is the sum of the losses due to weathering (λ_w) or degradation, including photodegradation (λ_{ad}). Thus,

$$\lambda_{al} = \lambda_w + \lambda_{ad}$$

The uptake from the soil component is given by:

$$U_s = \frac{B[1 - \exp(-\lambda_s\, t_b)]}{P\, \lambda_s}$$

where B = soil-plant uptake factor (unitless); λ_s = soil loss constant (s^{-1}); t_b = period of long-term buildup in soil (s); and P = density of soil in the root zone (kg/m²).

The soil loss constant is the sum of all soil losses due to leaching (λ_{sl}) or degradation (λ_{sd}), and thus:

$$\lambda_s = \lambda_{sl} + \lambda_{sd}$$

The estimation of the concentration of the pollutants in vegetation and forage crops is the first step in determining total exposure from the food chain. A determination of the con-

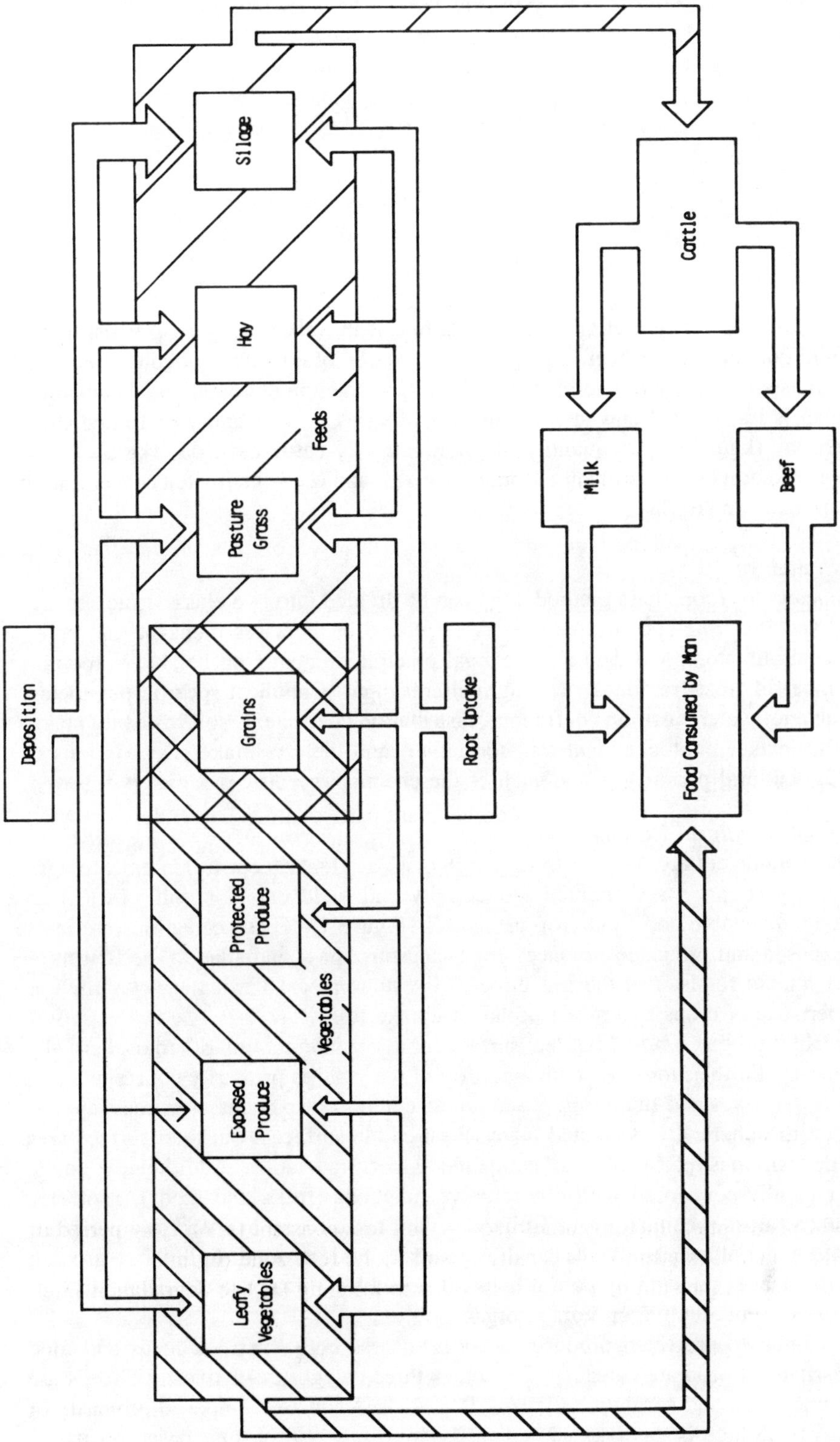

FIGURE 4. Categorization of vegetation transport characteristics and agricultural pathways.

centrations in beef and milk that result from ingestion by cattle of contaminated forage and grains is also necessary. Transport to beef and milk from ingested forage and grains is modeled via:

$$C_m = F_m(Q_F C_F + Q_G C_G)$$

and

$$C_f = F_f(Q_F C_F + Q_G C_G)$$

where C_m = estimated concentration of the contaminant in milk (μg/kg); C_f = estimated concentration of contaminant in beef (μg/kg); F_m = fraction of contaminant consumed each day which is transported and retained in milk (d/kg); F_f = fraction of contaminant consumed each day which is transported and retained in beef (d/kg); Q_F = quantity of forage eaten by cattle each day (kg/d); Q_G = quantity of grain eaten by cattle each day (kg/d); C_F = estimated concentration of contaminant in forage (μg/kg); and C_G = estimated concentration of contaminant in grain (μg/kg).

B. Input Parameters

Input parameters for food chain methodology can be divided into two classes: site specific and chemical specific. Site-specific parameters employed in the assessment include population data, local climatological data (i.e., annual precipitation, mixing height, evapotranspiration, number of frost-free days), and agricultural input. Chemical-specific parameters used are the octanol-water partition coefficient, distribution coefficient, soil-to-plant uptake, meat and milk uptake, photochemical transformation, and dietary intake. Descriptions of site-specific agricultural parameters and each of the chemical-specific parameters follow.

1. Site-Specific Agricultural Parameters

Agricultural parameters are derived from the 1974 U.S. Department of Commerce agricultural census by county[18] and include productivity and yield data for milk cows, beef cows, and seven vegetable and food crop categories (Figure 4). These categories are leafy vegetables, exposed and protected produce, grains, pasture, hay, and silage. The first three are classed as human foods, and the last three as livestock feeds. Grains are classified as both. Characteristics of crops that affect pollutant uptake follow.

Leafy vegetables present a broad flat leaf surface for direct interception of atmospherically deposited material. Furthermore, the edible portion of the plant is primarily concerned with vegetative growth (leaves and stems). Exposed produce items (snap beans, tomatoes, apples, etc.) intercept atmospherically deposited material on edible surfaces, but total surface area available for deposition is relatively small compared to leafy vegetables. Additionally, edible portions are typically concerned with reproductive functions (fruits and seeds). Protected produce items (potatoes, peanuts, citrus fruits, etc.) are not directly exposed to atmospherically deposited material because their growth is generally underground, or if above ground, the edible portions are protected by pods, shells, or nonedible skins or peels. Typically, the edible portions are reproductive or storage organs.

Grains are similar to protected produce, but their use as both livestock feeds and food for man necessitates a separate category. The other three categories of livestock feeds are pasture, hay, and (corn and sorghum) silage. All of these feeds are composed primarily of vegetative growth. Silage is categorized separately from hay and pasture based on its interception characteristics. Hay and pasture grass are separate because their residence times in the field are significantly different.

Table 11
OBSERVED AND CALCULATED LOG K_{ow} VALUES FOR ORGANIC CHEMICALS FROM INCINERATED WASTE STREAMS[22]

Chemical	Log K_{ow}	Ref.
1,1,1,2-Tetrachloroethane	3.05	21
Chloroform	1.96	25
	1.97	8
Ethylene dichloride	1.48	25
Hexachlorobutadiene	1.78	21
1,1,2,2-Tetrachloroethane	2.66	21
Hexachloroethane	4.62	21
Carbon tetrachloride	2.64	16
	2.83	25
	av = 2.74	
Hexachlorobenzene	5.23	13
	5.44	5
	5.50	6
	av = 5.39	
Trichloropropane	2.01	21
Trichloroethylene	2.29	25

2. Chemical-Specific Parameters

Chemical specific parameters describing food chain transport are necessary as input, but for most chemicals of interest, measurements are not available. Thus, it is necessary to calculate default values for these transfer coefficients from known physicochemical properties.

Fortunately, the physicochemical quantity known as the octanol/water partition coefficient (K_{ow}) is correlated with many of the parameters needed by the food chain transport model.[5,7,13] These parameters and their relationship to K_{ow} are discussed below.

a. Octanol/Water Partition Coefficient K_{ow}

The octanol/water partition coefficient is defined as the ratio of a chemical's concentration in the octanol phase to its concentration in the aqueous phase of a two-phase octanol/water system.[14] Values of K_{ow} represent the tendency of a chemical to partition itself between an organic phase and an aqueous phase and may be correlated to uptake of chemicals into biological systems. This parameter can be measured in the laboratory, although variability in measured values of K_{ow} for a given chemical may be affected by such factors as temperature, pH, time of mixing, purity of chemicals or solvents, etc.[13] Values reported in the literature for the organic chemicals considered in Chapter 9 are given in Table 11. If more than one value is reported, an average value is used. Leo's Fragment Constant Method[9] is utilized to derive log K_{ow} values for the organic chemicals for which there are no reported observations. These are also listed in Table 11. This is a method to estimate log K_{ow} from the chemical structure of the compound. A simple explanation of the method is available in Lyman et al.[14]

b. Distribution Coefficient K_d

The distribution coefficient, K_d, is a measure of the retention of a solute by the soil matrix and is represented by the ratio of the elemental concentration in the soil to the concentration in the solute, at equilibrium. This parameter is utilized in estimating soil concentrations which are then used to determine plant uptake from the soil. Following an extensive literature review of K_d values for compounds, an attempt can be made to correlate K_d to the reported

K_{ow} for each of the compounds.[3] By plotting log K_d vs. log K_{ow}, a straight line regression equation is derived.[5]

$$\log K_d = -0.90 + 0.521 (\log K_{ow})$$

K_d values estimated for the ten organic chemicals of interest in Chapter 9 are listed in Table 12.

c. Soil-to-Plant Uptake Parameters B_v and B_r

Root uptake of organic chemicals distributed in soil is described by the parameters B_v and B_r, representing soil-to-plant transfer coefficients for vegetative and nonvegetative portions of food crops, respectively. The parameters B_v and B_r are unitless as they represent the ratio of the pollutant concentrations in plants to the concentration in soil at time of harvest.

Baes[1] relates B_v to K_{ow}, deriving the following regression equation:

$$\log B_v = 2.71 - 0.62 (\log K_{ow})$$

Baes et al.[2] conclude that the reproductive parts of plants take up only a small fraction of a chemical incorporated into the vegetative portion of the same plant and thus assume that

$$B_r = 0.1 B_v$$

This is a relationship we believe to hold for organic compounds. Values of B_v and B_r estimated for the chemicals of interest in Chapter 9 are given in Table 12.

d. Meat and Milk Uptake Parameters F_f and F_m

Kenaga[12] shows a positive correlation between the bioconcentration factor (BCF) in beef fat (the ratio of the quantity of chemical found in beef fat to the quantity in the diet) and K_{ow}. Based on the average fat content of milk, various commonly ingested beef cuts, and the average consumption of feed per head of cattle,[1] derives the following regression equations:

$$\log F_f = -5.15 + 0.50 (\log K_{ow})$$
$$\log F_m = -6.13 + 0.50 (\log K_{ow})$$

Table 12 lists estimated values for F_f and F_m.

e. Photochemical Transformation λ_{ad}

Pollutants released by incineration facilities may undergo photochemical transformations which alter their chemical and physical nature. Photochemically induced changes in toxicity are not considered in this assessment. No photochemical degradation is assumed in estimating inhalation exposures since the atmospheric residence times over the area of concern (100 km) are relatively short (a matter of hours). However, pollutants deposited on food items may be photodegraded throughout the entire growing season(s). Chemical-specific values for λ_{ad} are used in estimating concentrations of pollutants in food items (Table 12).

f. Dietary Intake

Total dietary intake of major foodstuffs is based on United States Department of Agriculture (USDA) 1979 Household Survey data. The USDA study[23] estimates daily intake of foodstuffs

Table 12
ESTIMATED INPUT PARAMETERS FOR THE FOOD CHAIN MODEL[22]

Chemical	K_d (ml/g)	B_v	F_f (d/kg)	F_m (d/kg)	λ_{ad} (S^{-1})
1,1,1,2-Tetrachloroethane	4.23	6.3	2.4E − 4	2.5E − 5	2.0E − 8
Chloroform	1.12	30.0	6.9E − 5	7.1E − 6	6.5E − 8
Ethylene dichloride	0.62	60.3	3.9E − 5	4.1E − 6	5.5E − 6
Hexachlorobutadiene	0.90	40.4	5.5E − 5	5.8E − 6	1.1E − 5
1,1,2,2-Tetrachloroethane	2.63	11.0	1.5E − 4	1.6E − 5	1.2E − 7
Hexachloroethane	28.8	0.67	1.4E − 3	1.5E − 4	1.0E − 9
Carbon tetrachloride	2.90	10.0	1.7E − 4	1.7E − 5	1.0E − 9
Hexachlorobenzene	70.1	0.24	3.3E − 3	3.5E − 4	1.0E − 9
Trichloropropane	1.2	28.2	7.2E − 5	7.5E − 6	2.1E − 6
Trichloroethylene	1.7	18.7	9.9E − 5	1.0E − 5	1.1E − 6

Table 13
ESTIMATED ANNUAL AVERAGE DIETARY INTAKE (kg/YEAR)[22]

Foodstuff	Intake (kg/year)
Leafy vegetables	16
Exposed vegetables	35
Protected vegetables	64
Grains	75
Total dairy	112
Total meat	89

based on age and sex, and the EPA combines this data with the U.S. Bureau of Census data on population, age, and sex distributions to make average daily intake estimates. The values used for the assessment in Chapter 9 are listed in Table 13.

REFERENCES

1. **Baes, C. F., III,** Prediction of radionuclide K_d values from soil-plant concentration ratios, *Transl. Am. Nucl. Soc.*, 41, 53, 1982.
2. **Baes, C. F., Sharp, R. D., Sjoreen, A. L., and Shor, R. W.,** A Review and Analysis of Parameters for Assessing Transport of Environmentally Released Radionuclides through Agriculture, ORNL/TM-5786, Oak Ridge National Laboratory, Oak Ridge, TN, 1984.
3. **Baes, C. F. and Watson, A. P.,** personal communication, 1986.
4. **Bennett, B. G.,** Exposure of man to environmental PCB's — an exposure commitment assessment, *Sci. Total Environ.*, 29, 101, 1983.
5. **Briggs, G. G.,** Theoretical and experimental relationships between soil adsorption, octanol-water partition coefficients, water solubilities, bioconcentration factors and the parachor, *J. Agric. Food Chem.*, 29, 1050, 1981.
6. **Chiou, C. T. and Schmedding, D. W.,** Partitioning of organic compounds in octanol-water systems, *Environ. Sci. Technol.*, 16, 4, 1982.
7. **Chiou, C. T., Freed, V. H., Schmedding, D. W., and Kohnert, R. L.,** Partition coefficients and bioaccumulation of selected organic chemicals, *Environ. Sci. Technol.*, 11(5), 475, 1977.
8. **Hansch, C. and Anderson, S.,** The effect of intramolecular hydrophobic bonding on partition coefficients, *J. Org. Chem.*, 32, 2583, 1967.

9. **Hansch, C. and Leo, A.,** *Substituent Constants for Correlation Analysis in Chemistry and Biology,* John Wiley & Sons, New York, 1979, chap. 4.
10. **Harrington, E. S., Holton, G. A., and O'Donnell, F. R.,** An initial emission assessment of hazardous waste incineration facilities, in Proceedings, 1982 UCC-ND and GAT Environmental Protection Seminar, Oak Ridge National Laboratory, Oak Ridge, TN, 1982.
11. **Holton, G. A., Travis, C. C., Etnier, E. L., O'Donnell, F. R., Hetrick, D. M., and Dixon, E.,** Multiple Pathways Screening Level Assessment of a Hazardous Waste Incinerator Facility, ORNL/TM-8652, Oak Ridge National Laboratory, Oak Ridge, TN, September 1984.
12. **Kenaga, E. E.,** Correlation of bioconcentration factors of chemicals in aquatic and terrestrial organisms with their physical and chemical properties, *Environ. Sci. Technol.,* 14(5), 553, 1980.
13. **Kenaga, E. E. and Goring, C. A. I.,** Relationship between water solubility, soil sorption, octanol-water partitioning and concentration of chemicals in biota, in *Aquatic Toxicology Proceedings,* Eaton, J. G., Parrish, P. R., and Hendricks, A. C., Eds., ASTM STP 707, 1980, 78.
14. **Lyman, W. J., Reehl, W. F., and Rosenblatt, D. H., Eds.,** *Handbook of Chemical Property Estimation Methods, Environmental Behavior of Organic Compounds,* McGraw-Hill, New York, 1982.
15. **McBride, J. P., Moore, R. E., Witherspoon, J. P., and Blanco, R. E.,** Radiological impact of airborne effluents of coal and nuclear plants. Radiation doses from airborne effluents of a coal-fired plant may be greater than those from a nuclear plant, *Science,* 202, 1045, 1978.
16. **Neely, W. B., Branson, D. R., and Blau, G. E.,** Partition coefficients to measure bioconcentration potential of organic chemicals in fish, *Environ. Sci. Technol.,* 8(13), 1113, 1974.
17. **Raridon, R. J., Murphy, B. D., Culkowski, W. M., and Patterson, M. R.,** The Atmospheric Transport Model as Applied to Toxic Substances (ATM-TOX), ORNL/CSD-94, Oak Ridge National Laboratory, Oak Ridge, TN, November 1984.
18. **Shor, R. W., Baes, C. F., III, and Sharp, R. D.,** Agricultural Production in the United States by County: A Compilation of Information from the 1974 Census of Agriculture for Use in Terrestrial Food-Chain Transport and Assessment Models, ORNL/TM-5768, Oak Ridge National Laboratory, Oak Ridge, TN, 1982.
19. **Staley, L. J., Holton, G. A., O'Donnell, F. R., and Little, C. A.,** An assessment of emissions from a hazardous waste incinerator facility, Proceedings Eighth Annual Research Symposium on Land Disposal, Incineration, and Treatment of Hazardous Waste, Industrial Environmental Research Laboratory, U.S. Environmental Protection Agency, Cincinnati, OH, 1982.
20. **Travis, C. C., Baes, C. F., III, Barnthouse, L. W., Etnier, E. L., Holton, G. A., Murphy, B. D., Thompson, G. P., Suter, G. W., II, and Watson, A. P.,** Exposure Assessment Methodology and Reference Environments for Synfuel Risk Analysis, ORNL/TM-8762, Oak Ridge National Laboratory, Oak Ridge, TN, December 1983.
21. **Travis, C. C., Etnier, E. L, Holton, G. A., O'Donnell, F. R., Hetrick, D. M., Dixon, E., and Harrington, E. S.,** Inhalation Pathway Risk Assessment of Hazardous Waste Incineration Facilities, ORNL/TM-9096, Oak Ridge National Laboratory, Oak Ridge, TN, October 1984.
22. **Travis, C. C., Holton, G. A., Etnier, E. L., Cook, C., O'Donnell, F. R., Hetrick, D. M., and Dixon, E.,** Assessment of inhalation and ingestion population exposures from incinerated hazardous wastes, *Environ. Int.,* 12, 533, 1986.
23. **U.S. Department of Agriculture,** Agricultural Statistics, U.S. Government Printing Office, Washington, D.C., 1980.
24. *Background Document, Resource and Recovery Act, Subtitle C, Identification and Listing of Hazardous Waste,* Office of Solid Waste, U.S. Environmental Protection Agency, Washington, D.C., 1980.
25. **Valvani, S. C., Yalkowsky, S. H., and Roseman, T. J.,** Solubility and partitioning. IV. Aqueous solubility and octanol-water partition coefficients of liquid nonelectrolytes, *J. Pharmacol. Sci.,* 70(5), 502, 1981.
26. **Walsh, P. J.,** Principal investigator, *Health and Environmental Effects Document on Direct Coal Liquefaction — 1982, Vol. I, Assessment,* ORNL/TM-8624/V1, Oak Ridge National Laboratory, Oak Ridge, TN, August 1983.

INDEX

E

F

G

H

U

V

W